T0138608

Vehicle Feedback and Driver Situation Awareness

Human Factors in Road and Rail Transport
Series Editors Lisa Dorn, Gerald Matthews and Ian Glendon

Vehicle Feedback and Driver Situation Awareness

By
Guy H. Walker
Neville A. Stanton
Paul M. Salmon

CRC Press
Taylor & Francis Group
Boca Raton London New York

CRC Press is an imprint of the
Taylor & Francis Group, an **informa** business

CRC Press
Taylor & Francis Group
6000 Broken Sound Parkway NW, Suite 300
Boca Raton, FL 33487-2742

© 2018 by Taylor & Francis Group, LLC
CRC Press is an imprint of Taylor & Francis Group, an Informa business

Printed on acid-free paper

International Standard Book Number-13: 978-1-4724-2658-1 (Hardback)

Library of Congress Cataloging-in-Publication Data

Names: Walker, Guy, author. | Stanton, Neville A. (Neville Anthony), 1960, author. | Salmon, Paul M., author.
Title: Driver feedback in automotive engineering : the human factors of driver-vehicle interaction / Guy H. Walker, Neville A. Stanton, Paul M. Salmon.
Description: Boca Raton, FL : CRC Press/Taylor & Francis Group, 2018. | Series: Human factors in road and rail transport | "A CRC title, part of the Taylor & Francis imprint, a member of the Taylor & Francis Group, the academic division of T&F Informa plc." | Includes bibliographical references and index.
Identifiers: LCCN 2017057950| ISBN 9781472426581 (hardback : acid-free paper) | ISBN 9781315578163 (ebook)
Subjects: LCSH: Automobiles–Design and construction. | Human-machine systems. | Automobile driving–Human factors.
Classification: LCC TL250 .W348 2018 | DDC 629.2/31–dc23
LC record available at https://lccn.loc.gov/2017057950

Visit the Taylor & Francis Web site at
http://www.taylorandfrancis.com

and the CRC Press Web site at
http://www.crcpress.com

Contents

List of Figures

List of Tables

Acknowledgements

This book describes the authors' work which, over the past 15 years, has taken place in various institutions and under various funded projects. We would like to acknowledge the support of Heriot-Watt University, the University of Southampton, the University of the Sunshine Coast, Brunel University and Monash University. We would also like to acknowledge the important role of our sponsors, which have included Jaguar Cars, Ford Motor Company, the UK Engineering and Physical Sciences Research Council (EPSRC), the Australian Research Council (ARC), the Institute of Advanced Motorists (IAM) and the Carnegie Trust. The research reported in this book has been used within and compiled under the auspices of the EPSRC Centre for Sustainable Road Freight. It also includes research undertaken via a current ARC Discovery grant, and another provided by the Australian National Health and Medical Research Council. The majority of the work, however, was funded by the Hamilton Research Studentship. Brunel University's Department of Design, at which the lead author completed his PhD in 2002, was selected as the beneficiary of a substantial long-term donation bequeathed by a local entrepreneur and industrialist, J. D. Hamilton. The charitable donation is called the Ormsby Trust and it provided generously within the department, not least for the research reported in this book. The Ormsby Trustees, Rosemary David, Katrina McCrossan and Angela

Chiswell, are given especially prominent and grateful acknowledgment. Over these past 15 years, we have also worked with many friends and colleagues who have advanced their own research agendas using some of the same research facilities and equipment. We leave it to them to tell their own equally fascinating research stories but nonetheless would like to particularly acknowledge: Dr Dan Jenkins, Dr Ipshita Chowdhury, Dr Tara Kazi and, of course, all our many hundreds of experimental participants. All of them survived the on-road trials…even the motorcyclist who referred to our experimental course as the 'Runnymede Tourist Trophy'.

Glossary

ABS – Anti-lock braking system
ACC – Adaptive cruise control
ANOVA – Analysis of variance
bhp – Brake horse power
CDM – Critical decision method
CO_2 – Carbon dioxide
CVP – Concurrent verbal protocol
DSA – Distributed situational awareness
DSQ – Driving style questionnaire
ES – Effect size
GDTA – Goal directed task analysis
HTA – Hierarchical task analysis
HTAoD – Hierarchical task analysis of driving
IAM – Institute of Advanced Motorists
IoT – Internet of things
ITS – Intelligent transport systems
kW – Kilowatt
ns – Not significant
PN – Propositional network
SA – Situational awareness
SACRI – Situation awareness control room inventory
SAGAT – Situation awareness global assessment technique

SART – Situation awareness rating technique
SD – Standard deviation
SDT – Signal detection theory
SNA – Social network analysis
VP – Verbal protocol

About the Authors

Dr Guy H. Walker is an associate professor within the Centre for Sustainable Road Freight at Heriot-Watt University in Edinburgh. He lectures on transportation engineering and human factors and is the author/co-author of over 100 peer-reviewed journal articles and 13 books. He has been awarded the Chartered Institute of Ergonomics and Human Factors (CIEHF) President's Award for the practical application of ergonomics theory and Heriot-Watt's Graduate's Prize for inspirational teaching. Dr Walker has a BSc Honours degree in psychology from the University of Southampton, a PhD in human factors from Brunel University, is a fellow of the Higher Education Academy and a member of the Royal Society of Edinburgh's Young Academy of Scotland. His research has featured in the popular media, from national newspapers, TV and radio through to an appearance on the Discovery Channel. He has previously owned a Suzuki SC100 Whizzkid, two MkI Ford Fiesta XR2's, two Suzuki TL1000R superbikes and currently drives an Audi S1.

Professor Neville A. Stanton is a chartered psychologist, chartered ergonomist and chartered engineer and holds a chair in Human Factors Engineering in the Faculty of Engineering and the Environment at the University of Southampton. He has published over 300 peer-reviewed journal papers and 40 books on human factors and ergonomics. In 1998, he was awarded the Institution of Electrical Engineers Divisional Premium Award for a co-authored paper on engineering psychology and system safety. The Chartered Institute of Ergonomics and Human Factors awarded him the Sir Frederic Bartlett Award in 2012, the President's Award in 2008 and the Otto Edholm Award in 2001 for his original contribution to basic and applied ergonomics research. In 2007, the Royal Aeronautical Society awarded him the Hodgson Medal and Bronze Award, with colleagues, for their work on flight deck safety. He is also the recipient of the Vice Chancellor's Award for Postgraduate Research Supervisor of the Year in the Faculty of Engineering and the Environment at the University of Southampton. Professor Stanton is an associate editor of the *IEEE Transactions on Human-Machine Systems* and is on the editorial board of Theoretical Issues in Ergonomics Science. Professor Stanton is a Fellow and chartered occupational psychologist registered with the British Psychological Society, a Fellow and chartered ergonomist registered with the Institute of Ergonomics and Human Factors Society, and a chartered engineer registered with the Institution of Engineering and Technology. He has a BSc in occupational psychology from the University of Hull, an MPhil in applied psychology, a PhD in human factors engineering from Aston University in Birmingham and a DSc in human factors engineering awarded by the University of Southampton. He has previously owned a 1978 Mini 1000, a Mark 1 Ford Escort, a Saab 900, a Yamaha FS1E, a Honda 400/4, a Honda CB750K and a Honda CBR1100XX Super Blackbird. He now owns a Toyota Land Cruiser, Mazda MX5 and Triumph Bonneville T100.

Dr Paul M. Salmon is a professor in human factors and leader of the University of the Sunshine Coast Accident Research (USCAR) team at the University of the Sunshine Coast. He holds an Australian Research Council Future Fellowship in the area of road safety and has over 13 years of experience in applied human factors research in a number of domains, including military, aviation, road and rail transport. Paul has co-authored 10 books, over 90 peer-reviewed journal articles and numerous conference articles and book chapters. Paul has received various accolades for his research to date, including the 2007 Royal Aeronautical Society Hodgson Prize for best paper and the 2008 Ergonomics Society's President's Medal. Paul was also recently named as one of three finalists in the 2011 Scopus Young Australian Researcher of the Year Award. He has previously owned a Fiat Cinquecento Sporting, two MkIII Ford Escort Cabriolets and currently drives a Jeep Wrangler...often on the beach.

1

INTRODUCTION

1.1 When 70 mph Really Felt Like It

A potentially troubling aspect of modern vehicle design – some would argue – is a trend towards isolating the driver and reducing vehicle feedback, usually in the name of comfort and refinement but increasingly because of automation (e.g. Loasby, 1995). It is an interesting thesis. There can be little doubt cars have become more civilised over the years yet, despite this, the consequences on driver behaviour remain to a large extent anecdotal. Readers will have heard such anecdotes for themselves. They usually take the form of drivers of a certain age recalling their first cars from the 1970s or 1980s, in which 'doing 70 mph really felt like it'. The question is: do such anecdotes reflect a bigger, more significant issue that could be better understood? Related questions have been explored in other domains such as aviation, where the change to 'fly-by-wire', for example, did indeed bring about some occasionally serious performance issues that were not anticipated (e.g. Sarter and Woods, 1997; Field and Harris, 1998). Despite some clear parallels automotive systems have been left relatively unexamined (MacGregor and Slovic, 1989; Stanton and Marsden, 1996). The research described in this monograph aims to explore precisely these issues from a Human Factors perspective.

What has happened to car design to make anecdotes about it even mildly interesting or relevant? Well, cars have certainly become faster, more powerful, heavier and more sophisticated. These changes are driven in large part by legislation. In particular, car makers are now subject to financial penalties if their fleet average emissions exceed legislated limits (Martin et al., 2017). This has meant the conventional powertrain – the internal combustion engine – has undergone significant development in order to become cleaner while at the same time meeting customer demands for performance. A small historical diversion serves to illustrate just how far vehicle technology has actually travelled.

Case one is extreme: the Porsche 911. In 1979 the new Turbo model, with its six cylinder 3299 cc engine, developed what road testers of the day referred to as a 'staggering' 300 bhp and 303 lb ft. of torque. They wrote that: "the acceleration is startling [...] After only 2.2 sec, the car was travelling at 30 mph, 40 mph came up in 2.7 sec, 50 in 3.7 sec, 60 in 5.3 sec. This last figure [...] break[s] new ground, making the Turbo the quickest production car we (and as far as we know, anyone else) have ever tested" (Motor, 1979, p. 225). The car also went on to record "the highest top speed we have ever measured for a production model" (p. 255), judging that a mean speed of 160.1 mph "should satisfy most owners".

With that hyperbole in mind, fast-forward now to 2018 and the current Volkswagen Golf hatchback. In its top specification the Golf now has more power (306 bhp) developed from an engine 40% smaller and with two fewer cylinders. The Golf can accelerate from 0–60 mph faster than the Porsche (4.9 seconds), has a higher theoretical top speed (it is electronically limited to 155 mph, but a variant without this achieved 165 mph), but even more remarkably, has well under half the fuel consumption potential (40.9 mpg on the combined cycle versus 15.9 mpg). Put another way, it has taken approximately forty years for performance that was once the preserve of supercars – and only the fastest of those – to arrive in a small family hatchback (Figure 1.1). Frankly, it would be surprising if such a change had not had an effect on driver behaviour and cognition.

Case two looks at this progress from a different angle, selecting an archetypal UK car in which '70 mph *really* felt like it': the Austin Mini. For readers in other territories have in mind an equally small, basic and slow 'first car', such as a Geo Metro, Volkswagen Beatle, or Datsun Cherry. Originally introduced in 1959 the Mini had a cast iron pushrod 'A-Series' engine, the most common version displacing 998 cc. In this form it generated just 39 bhp, 51.4 lb ft. of torque, and endowed the Mini with a 0–60 mph time of approximately 19 seconds. In the late 1970s, when some readers may have been purchasing a Mini as their first car, contemporary road testers were describing it as: "[...] slow, at times painfully so, [it] struggles to make the legal limit. For long distance motorway work it is awful"

1979 Porsche 911 Turbo 2017 Volkswagen Golf R

Image: Wikimedia Commons

Image: Wikimedia Commons

"The acceleration is startling. For our MIRA [test track] starts, we were dropping the clutch abruptly at 6,000 rpm, leaving two expensive black lines of rubber on the road for a considerable distance. After only 2.2 sec, the car was travelling at 30 mph, 40 mph came up in 2.7 sec, 50 in 3.7 sec, 60 in 5.3 sec. This last figure and the 0–100 mph time of 12.3 sec also break new ground, making the Turbo the quickest production car we (and as far as we know, anyone else) have ever tested. The standing quarter mile is covered in 13.4 sec, passing the post at 104 mph; if this is not quite in the same league as a dragster, it is certainly enough to blow off almost anything else on the public road."

Motor, 1979, p. 225

"It was a streaming wet day at MIRA when we tested the DSG-equipped variant of the Golf R, but the natural advantage of four-wheel drive came to the fore during the acceleration runs. Even two up, full of fuel and in the wet, it was a 4.8 sec 0–60 mph car in our hands."

Autocar, 2017, https://www.autocar.co.uk/car-review/volkswagen/golf-r

Figure 1.1 1979 Porsche 911 Turbo versus 2018 Volkswagen Golf.

(Motor, 1979, p. 146). One of the reasons 70 mph felt very much like 70 mph was quite simply that noise levels "are unbearable above 50 mph. Driving at an indicated 70 mph on motorways gave two testers headaches and any thoughts of listening to the radio are ruled out. […]" (p. 147). Compounding matters was that "on big bumps the car – and its occupants – pitch about uncomfortably. […]" (p. 147). Summing up the experience the reviewers stated: "The appalling lack of refinement in such major areas as noise and ride comfort, coupled with the very poor performance, put the car way behind many competitors" (p. 147). Except, perhaps, in one important area: it *felt* fast…

With that thought in mind, fast-forward again to the present day. The 'same' vehicle, a base-model 2018 Mini One, fitted with its new BMW B38 engine (an all-aluminium turbocharged 1198 cc three cylinder) develops nearly three times more power (102 hp) and torque (133 lb ft.). It can accelerate from 0–60 mph in half the time (9.9 seconds) despite being nearly twice as heavy (1150 vs. 630 kg). In terms of fuel consumption the differences are even more acute. The newer car has 70.6 mpg fuel economy potential which is over double the older car's 33.1 mpg 'touring' figure (the most favourable quoted by *Autocar*). Remember, this is despite nearly double the weight and triple the power. The most startling features, however, are more hidden. Forty years ago the Mini – like virtually all cars – had no catalytic converter, no direct injection, no engine management, no turbocharger, no variable valve lift and timing, no automatic start-stop, no drive-by-wire, no CAN-bus, no driver display showing fuel economy, no independent engine cooling circuits and no low friction engine technology. On the contrary, it ran on leaded four-star petrol, had only four gears, and its CO_2 emissions were approximately 198 g/km. To put that in context, if the 1978 Mini 1000 were a new car, and despite it only having a small 1-litre engine, it would be in the UK government's third highest road tax band and it would, at the time of writing, cost the owner a rather off-putting £1200 for the first year and £140 thereafter. Cars to be found in this bracket today include the supercharged version of the Jaguar XE, the Toyota Land Cruiser, the AMG Mercedes S-Class saloon, and an eleven-seat Ford Transit minibus. The modern Mini One, on the other hand, has emissions of just 109 g/km. Even the present day Porsche 911 Turbo has better CO_2 emissions than the 1978 Mini (Figure 1.2).

These amusing case studies do in fact bear out wider trends in vehicle design, as extensively researched by colleagues within the UK's Centre for Sustainable Road Freight (www.csrf.ac.uk). Using a large and unique dataset of 35,000 distinct vehicle models, Martin et al. (2017) reveal a 22% reduction in emissions for turbocharged petrol powered vehicles (such as Minis and Porsche 911s) and a dramatic

Image: Wikimedia Commons

Figure 1.2   'Classic' Mini versus 'Modern' Mini.

38% increase in so-called 'power density', all in the 10 years spanning 2001–2011. The anecdotes about Volkswagen Golf GTis and Porsche 911s do indeed bear out wider trends. The Porsche 911 Turbo was exotic in 1978 precisely because it used a turbocharger. Well over 40% of all light duty vehicles sold in the EU now have one. In the early 1970s Porsche had successfully introduced turbocharging in their Le Mans racing cars and the years following became known as the 'Turbo Era' in Formula 1. Turbocharging in this context had little to do with emissions and fuel efficiency as it does today. One of the prominent technical barriers for 1980s Formula 1 cars was how quickly their 1000 bhp turbo engines consumed the 195-litre fuel allowance. Today, turbochargers, in concert with advanced fuel injection technologies, "allow vehicle performance to be decoupled from the size of the engine, thus increasing engine power while reducing engine displacement volume" (Martin et al., 2017, p. 13). The VW Golf/Porsche 911 case study is an extreme example of this 'decoupling' and shows how turbocharging enables a small engine to match the performance of a big one (Figure 1.3).

If turbocharging a 2-litre VW Golf engine can make it match the performance of a 3.3-litre Porsche one, it follows that turbocharging a 1-litre engine will make it match the performance of a 1.6-litre one, while at the same time conferring most of the efficiency and emissions benefits of the original 1-litre engine. This is what a recent trend by

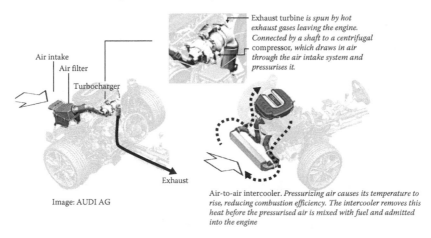

Air intake
Air filter
Turbocharger

Exhaust turbine *is spun by hot exhaust gases leaving the engine. Connected by a shaft to a centrifugal* compressor, *which draws in air through the air intake system and pressurises it.*

Exhaust

Image: AUDI AG

Air-to-air intercooler. *Pressurizing air causes its temperature to rise, reducing combustion efficiency. The intercooler removes this heat before the pressurised air is mixed with fuel and admitted into the engine*

Figure 1.3 Turbocharging is one of the significant technologies that enable vehicle performance to be decoupled from engine size: a smaller engine can be made to deliver equivalent power to a larger one.

the name of 'downsizing' achieves, a trend which is clearly discernible (Martin et al., 2017). The number of engine cylinders across the passenger vehicle fleet is declining at the rate of approximately 5% per year (Martin et al., 2017) and allied technologies such as turbocharging and direct injection are increasing likewise. What results are small, high-tech, highly tuned engines that weigh under 100 kg and occupy the footprint of an A4 sheet of paper. While the accent is firmly on efficiency and emissions, the favourable performance enhancing benefits of technologies like turbocharging remain. Some of the latest breed of 'downsized' engines have specific outputs in excess of 120 bhp/litre, turbocharger boost pressures double that used in the 1979 Porsche 911 turbo, and not that far off the kinds of pressures used in the Formula 1 'turbo era', at the time considered extreme. Why is this important? Because in order to make these unruly 'racing technologies' suitable for road use the focus has had to switch to 'usability'. This is not a straightforward engineering question. Consider this:

> *"the sound of two-cylinder* [downsized] *engines yields half the perceived engine speed of an equivalent four-cylinder engine at the same engine speed. As a result when driving, the two-cylinder* [downsized] *engine would be shifted to higher gears much later, diminishing the expected fuel savings"* (Sontacchi et al., 2016, p. 13).

What we have here is something akin to a classic 'irony of automation'. A technology aimed at improving fuel consumption yields the opposite effect to that intended. In this particular study:

> *"the optimal theoretical gear change* [for fuel economy] *should happen at around 2000 revolutions per minute (rpm). Studies under practical conditions show for an examined two-cylinder engine* [...] *that the typical gear change occur almost at 4000 rpm"* (p. 14).

With an acoustic device implemented that could imitate the sound of a larger four-cylinder engine – and with no other changes made to the vehicle – drivers shifted gear sooner. More interesting was drivers rated throttle response and engine/drivetrain smoothness as being significantly better, even though it was identical. Clearly, then, there is not only a disconnect between vehicle performance and engine size (e.g. Martin et al., 2017, p. 13) but between engineering properties of automotive systems and psychological ones. A disconnect that can have significant impacts which are explored in this book.

1.2 Vehicle Dynamics

The changes so far described are significant enough but even these ignore the fact that fossil-fuel powered reciprocal piston engines are, to say the least, a mature technology, one that, in essence, has been with us since the dawn of the automobile itself. A paradigm shift to new electric and hybrid drivetrains and full vehicle automation is well under way, with even greater changes to vehicle design in progress. Human factors research of the kind presented in this book is therefore timely. What is in prospect for vehicle designers is the ability to yield meaningful reductions in fuel use and emissions, through a better understanding of how complex vehicle systems mediate actual driver behaviour. There is also the not insignificant ability to compete on 'the driver experience'. This is an important factor given the legislative constraints driving the convergence of existing automotive technologies and the need to differentiate in other areas. There is also the critical need to ensure disruptive new technologies (such as EVs) deliver what drivers want. This is so that unexpected secondary effects do not diminish the primary benefits of adopting a new technology in the first place. That would indeed be ironic.

What connects these powerful changes in vehicle design to the driver is feedback. Feedback is "...sending back to the user information about what action has actually been done, what result has been accomplished..." (Norman, 2013, p. 27) or knowledge of results (Annett and Kay, 1957). Driving is a manual control task. Drivers operate the vehicle's 'control inceptors' (the pedals, steering wheel etc.) by moving their arms and legs, and this movement, in turn, is a product of human cognition. Chapter 3 goes on to explain in detail how this manual control task is part of a 'tracking loop' in which drivers have to perceive the state of their environment and issue 'command inputs' in order to neutralise 'errors' in vehicle speed or direction. Vehicle feedback is the relationship between the human inputs to the vehicle controls, the resulting movement of the control, and the resultant movement of the vehicle in response to the control input.

The vehicle's response to these demands is in turn fed back to the driver through the changing weight or effort needed to manipulate the controls, the vehicle motion in response to control inputs, changes in engine sounds, and a myriad other sensations resulting from the driver-vehicle-road interaction. Some of these sensations, as we saw above with the two cylinder versus four cylinder engine noise, give rise to significantly different driver behaviours.

By dealing with issues of tracking loops, manual control tasks and feedback, we stray into a topic normally occupied by vehicle dynamics. Vehicle dynamics is the study of how the different elements of vehicle body, propulsion, guidance and suspension devices, and the road itself, interact (Popp and Schiehlen, 1993). One of the purposes of this book is to work at the cross-disciplinary interface between the human sciences and this often rather august engineering field. This again is not as straightforward as it seems.

The first issue is one of ontology. This describes the differing world-views underlying the disciplines of vehicle dynamics (founded on classical dynamics) and human factors (founded on the behavioural sciences). The former relies on a detectable physical reality (positivist) while the latter often accepts a more partial detection of largely cognitive realities (interpretivist). These differences are particularly manifest in the underlying 'modelling languages' used by the two disciplines. In the behavioural sciences strict axioms and mathematical formulations are not suitable for the often-probabilistic phenomena under investigation. In the formal study of vehicle dynamics, on the other hand, "we must rely on clear definitions and model formulations, and then on a rigorous mathematical analysis. We must, indeed, 'formulate' the problem at hand by means of mathematical formulæ. There is no way out" (Guiggiani, 2014, p. 1).

Of course, no one is disputing the value of mathematical formulations, nor what the mathematical study of vehicle dynamics has contributed to vehicle design. After all, it has given us 1979 Porsche 911 Turbo performance in a Golf GTi. It is, however, fair to say this type of mathematical formulation occupies a region within a wider 'problem space' in which other ontologies co-exist. It is not the singular explanatory conceptual and modelling language, no matter how convenient it may be to regard it as such. The same progress in vehicle dynamics which has taken us from the 1970s era Austin Mini to today's version has exposed features of the vehicle/driver system that cannot be easily formulated yet can no longer be ignored. This brings us to the second point. There is an element in the vehicle dynamics control loop often conspicuous by its absence (e.g. Guiggiani, 2014; Popp and Schiehlen, 1993; Pacejka and Besselink, 2012), an element that does not lend itself at all well to the strict formalism of rigorous mathematical analysis: a human.

The engine downsizing example given above is a good one. The strictly formulated vehicle/road system does not need engine noise of any sort. It is simply a by-product of the combustion process. By including the driver in the control loop, however, everything changes. Suddenly the performance of the vehicle 'system' on critical performance variables such as fuel use and emissions is considerably altered, and in a less desirable way. This is a theme picked up in Chapter 2 where we ask, 'Who is the vehicle feedback for?' and we discover not just the ways in which humans put this feedback to use, but how incredibly sensitive they are to it. Vehicle dynamists have studied this before and the results are surprising. Normal drivers are able to detect changes in vehicle handling of a magnitude equating roughly to "...the difference in feel of a medium-size saloon car with and without a fairly heavy passenger in the rear seat" (Joy and Hartley, 1953/4, p. 263). Human factors researchers, meanwhile, show that having detected these subtle feedback cues, drivers use them for all manner of purposes, many of which are unexpected. In the case of engine noise, for example, it has been found that "drivers who received the quieter internal car noise [...] chose to drive faster than those who received louder car noises" (Horswill and McKenna, 1999). Remember, this is a form of vehicle feedback the vehicle itself does not need. The paradox, which we will explore throughout this book, is that this level of sensitivity to vehicle feedback, and the diversity of driver responses to it, stands in contrast to the dramatically increasing power, authority and autonomy of current and future vehicle systems. The probability of unexpected behavioural side effects occurring when well intentioned automotive technologies come into contact with drivers is therefore high.

This creates opportunities as well as hazards. Pioneering work by other colleagues within the Centre for Sustainable Road Freight (e.g. Nash et al., 2016; Na and Cole, 2015; Odhams and Cole, 2014) is reaching out across this cross-disciplinary divide with some extremely novel ways of incorporating 'the human element' in the vehicle dynamics system. In doing so they are capitalising on some of these inherent opportunities. There is much more work to be done, though. If vehicle dynamists are reaching out across the 'problem space' then this book, in its turn, represents an attempt by the behavioural sciences to reach out in the other direction. Chapter 3, for example, casts the car as a human factors artefact and asks, using concepts we hope human factors practitioners and vehicle dynamists alike will recognise, what sort of artefact it is. Crossing the disciplinary divide is not a task that can be completed in just one book. We hope, nevertheless, to shed some light on what might be happening within that anonymous black box labelled 'driver' in the all too rare cases of it appearing in vehicle dynamics control diagrams.

1.3 Driver Situation Awareness

Where the formal study of vehicle dynamics might struggle is an area that human factors can assist: human cognition. Human factors lifts the lid of the control loop element labelled 'human' or 'driver' through the examination of mental processes. These lie at the heart of cognitive psychology on which, to a very large extent, the study of human factors is founded. Ulric Neisser, an early pioneer in the field, refers to cognition as "all processes by which the sensory input is transformed, reduced, elaborated, stored, recovered, and used" (1976, p. 4). Cognitive psychology has historical roots in cybernetics, information theory and other allied concepts so, despite differences in underlying ontologies, vehicle dynamics and cognitive psychology/human factors have more in common than might at first be appreciated. At the very least, both make extensive use of control theory concepts and control theory inspired diagrams. Chapter 3 shows how some of these can be integrated. Where human factors begins not to 'compete' with vehicle dynamics in terms of strict formalism are the specific cognitive processes, such as 'situation awareness', through which sensory input is transformed, reduced, elaborated and used. To date there is not a mathematical model of the human mind and human factors has to satisfy itself with causal links of a more probabilistic nature. This does not diminish their usefulness. This volume of work selects one of the most practically expedient and 'useful' human factors concepts currently in existence: situation awareness.

The importance of situation awareness (SA) within driving cannot be understated. A key component of driving is knowing about the vehicle's current position in relation to its destination and the relative positions and behaviour of other vehicles and hazards, and also knowing how these critical variables are likely to change in the near future (Gugerty, 1997; Sukthankar, 1997). Moment-to-moment knowledge of this sort enables effective decisions to be made in real time and for the driver to be 'tightly coupled to the dynamics of [their] environment' (Moray, 2004). The link between vehicle feedback and driver SA can be considered all the more important given that poor SA is a greater cause of accidents than improper speed or driving technique (Gugerty, 1997). The central irony explored in this book is that modern trends in automotive design often appear – arguably – to be diminishing the level and type of vehicle feedback available to the driver, and thus potentially diminishing their SA. It is for reasons like this that SA is one of the most keenly studied topics in human factors (Wickens, 2008; Salmon and Stanton, 2013; Stanton et al., 2010). As a concept it provides researchers and practitioners with various models and methods to either describe what driver SA comprises, to determine how drivers develop SA, or to assess the quality of SA during driving task performance (Salmon and Stanton, 2013). It should also provide explanations for what happens when SA is lost, and how it affects performance when it is gained (Stanton et al., 2015).

Unusually for a human factors concept it has entered the mainstream lexicon. SA is used in many contexts and professions to refer interchangeably to information that resides in people's heads, minds (Fracker, 1991; Sarter and Woods, 1991; Endsley, 1995) or even brains (Endsley, 2015); as something which exists in the world, in displays or other environmental features (e.g. Ackerman, 1998, 2005); as something which is an emergent property of people and their environment (Stanton et al. 2006, 2009a, 2010); or as a form of distributed cognition (e.g. Hutchins, 1995a, b). SA has been explored in many areas, ranging from military settings (e.g. Endsley, 1995; Salmon et al., 2009; Stanton et al., 2006; Stanton, 2014; Stewart et al., 2008), sport

(Bourbousson et al., 2011; James and Patrick, 2004; Macquet and Stanton, 2014; Neville and Salmon, 2016), health care and medicine (Bleakley et al., 2013; Fioratou et al., 2010; Hazlehurst et al., 2007; Schulz et al., 2013), process control (Salmon et al., 2008; Sneddon et al., 2015; Stanton et al., 2009b), the emergency services (Seppanen et al., 2013; Blandford and Wong, 2005) and, of course, driving (e.g. Ma and Kaber, 2007; Golightly et al., 2010, 2013; Salmon et al., 2014; Walker et al., 2008, 2009). The papers which deal with the concept are among the top cited in the human factors discipline (e.g. Lee et al., 2005) and the term SA is one of the most widely used (Patrick and Morgan, 2010). Despite its prevalent use in many areas our understanding of SA is still in development (Flach, 1995; Dekker, 2015) and a feature of this book is how the study of driving allows a range of different SA 'lenses' to be inserted across the frame, each revealing different findings.

The definition of SA has changed and developed over the past 25 years, reflecting a change in foci of the human factors discipline as well as the facets of SA. Early definitions of SA focused on the individual person, as defined by Endsley:

"Situational awareness is the perception of the elements in the environment within a volume of time and space, the comprehension of their meaning and a projection of their status in the near future" (Endsley, 1995, p. 36).

As research interest in SA grew, so it expanded from individuals to teams. Definitions of SA came to reflect this also:

"the shared understanding of a situation among team members at one point in time" (Salas et al., 1995, p. 131).

Contemporary research into systems ergonomics has led to an interest in applying SA across wider sociotechnical systems. Again this has led to new definitions, such as:

"activated knowledge for a specific task within a system…[and] the use of appropriate knowledge (held by individuals, captured by devices, etc.) which relates to the state of the environment and the changes as the situation develops" (Stanton et al., 2006, p. 1291).

In driving, for example, one could study the awareness of individual drivers as we do in Chapters 2 and 5. Or it could be expanded to include other road users, as in Chapter 7 where we examine not just car drivers but other vehicle users that are being interacted with on the road. The analysis could also include more advanced forms of driver assistance systems as in Chapter 9. After all, vehicle automation is a 'system agent' in the control loop that, to some extent, has to be aware of its situation in order to function, or at the very least, help human agents like the driver do so. It is possible to go even further than this, to consider the wider, city-scale system of agents and actors within which the vehicle is operating. Issues like these lie at the heart of connected and autonomous vehicle debates. In Chapter 9, therefore, we insert this lens across the SA frame and ask whether improvements in individual driver SA lead to collective improvements at the level of entire street networks. The results are intriguing and surprising.

Not only is consideration of human and non-human 'agent' awareness becoming increasingly important given technological advances such as artificial intelligence and advanced automation, but so too is the notion of multiple agents cooperating over time in order to remain coupled to the dynamics of their environment. Issues like these lie at the heart of sociotechnical systems (STS) approaches (Walker et al., 2009). From this perspective, rather than simply defining the characteristics of individual elements and seeking to combine these, one has to study the system at the appropriate level, be it individuals, teams or entire systems. We will use this idea of levels of analysis to frame the analyses of vehicle feedback and driver SA in the following chapters.

1.4 Lapping the Driver Feedback Test Track

There are currently in excess of 1 billion cars in the world, one for every seven people, including children and infants. Love it or loathe it, the motoring program *Top Gear* holds the Guinness World Record for the world's most widely watched factual TV programme. Cars are ubiquitous. So if nothing else we hope this book is one that many readers can relate to through their own experiences of driving and the kinds of vehicles they themselves have owned. With that in mind, the time has come to perform a sighting lap around the *Top Gear* test track and introduce the topics and studies to be covered in this monograph.

Chapter 2 begins by asking a fundamental question: what exactly has happened to car design and, to put it bluntly, does it matter? The study reported in this chapter occurred at a unique period in time. The vehicle fleet still contained a useful proportion of older vehicles which lacked certain technologies, such as power steering, which are now almost universal. This enabled us to assess the SA of drivers of older vehicles (mean design age of 1983) compared to drivers of newer vehicles (with a mean design age of 1998). In answer to the quote at the beginning of this chapter, that drivers have become isolated from their environment, the answer appears to be 'yes'. The 'quantity' of SA for drivers of the newer vehicles is certainly lower than that for drivers of the older vehicles and this, in turn, points towards deeper, more fundamental issues. To explore these issues it becomes necessary in Chapter 3 to scrutinise 'the car' from a human factors perspective. To step back and ask what sort of human factors artefact it actually is. It is certainly one of the most complex available for consumers to use and buy yet, despite this, humans seem remarkably well adapted to it. The human factors properties of vehicle controls help to explain why this is so, but also how those same properties might be unfavourably altered by otherwise well-intentioned technologies. This chapter uses the language of tracking tasks, control loops and transfer functions, and is a point around which the disciplines of human factors and vehicle dynamics can hopefully gather.

Having identified driver feedback as being important, and that vehicle technologies may alter some of the fundamental properties of vehicle controls and, by extension, driver SA, it then becomes natural to ask a further question: what is the driving task and what are the attendant SA requirements? This is the task to which Chapter 4 turns. Formal human factors methods, such as task analysis, exist with which to systematically and exhaustively decompose the driving task for the purposes of identifying drivers' information needs. This chapter enables us to pause and consider the oft-quoted epithet that driving is a predominantly visual task. Is this really the case? How easy would a vehicle be to drive if there were no other sensations apart from visual input? In other words, if the driver were in a state of complete non-visual isolation? At the very least it would be a very different driving experience. The exhaustive SA requirements analysis presented in Chapter 4 shows driving is perhaps more correctly viewed as a multi-modal task, one in which there are multiple sources and modalities of feedback, connected to different facets of the driving task, in complex ways. In a practical sense this enables vehicle designers to understand what drivers' information needs are and to innovate ways of supporting them. Indeed, by not doing so one is confronted by a notable irony which forms the topic of Chapter 5. Stated simply there is a risk we design out precisely the feedback drivers want and need in favour (perhaps) of focusing on yet more visually dominant information sources from arbitrary and disconnected forms of technology. Taken to its limit this argument would run to a future vehicle that places drivers in a sensory – possibly even 'existential' – bubble, isolated from the vehicle's continuous auditory and tactile 'chatter' and, as a result, creates a driver who is less able to keep track of critical variables, error prone, and not able to intervene in cases where automation might require it. In this chapter we make use of a driving simulator to place drivers in precisely this sensory bubble and undertake a number of multi-modal vehicle feedback comparisons. Not surprisingly the results show feedback does increase driver SA, thus the irony of accidentally 'designing it out' is justified. More interestingly is when drivers themselves were asked simply how situationally aware they thought they were, there was no difference across conditions. In other words, drivers' awareness of their own situational awareness was extremely poor. Remember, SA is a greater cause of accidents than improper speed or improper driving technique (Gugerty, 1997). If drivers themselves cannot communicate the state of their own SA directly, then it falls to vehicle designers to do it for them, using human factors methods and insights to do so.

We are now approaching the 'Hammerhead' on the *Top Gear* test track, an aggressive left–right transition serving as a metaphor for our next two chapters which examine motorcycles. The results of Chapter 5 were diffused among the popular media (Figure 1.4) and generated lots of discussion. One such discussion was with a vehicle designer at Ferrari where the relative merits of a classic W-Class Mercedes were compared to more recent models, and the decision to compare motorcycles with cars was taken. From a purely experimental point of view it was clear motorcycles represented a particularly extreme manipulation of vehicle feedback. Motorcycles have no 'power assisted' controls of any sort; they do not even have a vehicle

Monday, October 15, 2007 METRO 15

Drivers 'at risk from quiet cars'

BY JOEL TAYLOR

MOTORISTS could be putting lives in danger because modern cars are almost 'too well' designed.

The lack of engine and road noise means drivers of newer cars are getting less feedback about road situations than when they drive older vehicles, research sho...

isolated in their cars from the outside environment, and less aware of what is happening around them, reveals the study from Brunel University's driving

...id: 'Our research ... need to balance with user-centric vers have appro- e feedback; uch as engine ise, keeps us on on the road.' feedback is esign innova- re of driving y, potentially

Warning over design of 'silent killer' cars

Manchester Evening News, 15th Oct 2007

Figure 1.4 The research into vehicle feedback and driver SA led to a variety of 'interesting' headlines.

body to shelter the 'occupant' from the external environment. Perhaps more importantly they had never before been subject to this kind of human factors analysis. Chapter 6, therefore, proceeds to describe an on road study in which car and motorcycle feedback and driver/rider SA were compared and the stark differences made manifest. Motorcyclists have considerably 'more' and 'different' SA compared to car drivers. Chapter 7 then proceeds to slide a team/systems SA lens across the analytical frame. Do the manifest differences shown in Chapter 6 have an effect on the cognitive compatibility of these two road users? In other words, does the mental representation of the situation developed from the feedback emanating from these two very different vehicle types affect the way they interact? After all, they both share the same operating environment. If there are differences, are they mutually compatible or incompatible? The findings of this on-road study reveal considerable differences in how motorcyclists and drivers perceive the same environment. Some of these differences are mutually reinforcing but others are not. Areas of incompatible SA happen to occur at sites/locations we know from previous research are particularly accident prone. A number of practical interventions are driven off these findings, one of which is training.

Chapter 8 explores the impacts on driver SA of advanced driver training. This is a form of driver training which makes explicit the need to take information from the environment and give it to other road users. It is also a training scheme founded on an established 'heuristic' based on five phases, with 'phase 1: information' overlaying all others. The driver training topic allows the 'systems SA' lens to be slid across the frame. This is because some of the shortcomings of adopting an individualistic view of SA become apparent when scrutinising the notion of 'expertise'. One in particular is the notion that the better the driver's 'mental theory' or SA of a situation is, the more abstract and simplified it will be, and the less likely it will actually look like the situation being perceived. Indeed, the paradoxes begin to accumulate when it is realised many individualistic approaches to SA, whilst useful for some research questions, are less useful when a 'literal' model of a situation is more likely to be held by novices rather than experts. To answer the more direct question about whether driver SA can be trained, Chapter 8 finds in the affirmative: following training drivers show an increase in the number of situational elements they are aware of, greater levels of interconnectivity between them, favourable changes in the criticality of certain elements and, above all, favourable changes in actual behaviour. In other words, advanced drivers are creating better situations to be aware of. The interesting methodological aspect to note is the opposite conclusion would be drawn had the systems lens not been slid across the analytical frame.

Picking up speed we drop our nearside wheels off the test track and hit the 'Follow Through' with Chapter 9. A consistent theme (read: 'concern') among the user-centred professionals we work with at vehicle manufacturers is that new display technologies will proliferate in an unstructured and random way. Unchecked they will gradually fill up the available dashboard space and overwhelm the driver who is expected to make sense of it all. It is a truism to say that, in future, the vehicle will have to provide new and different types of feedback to drivers in order to support equally new and different types of driving tasks. The question now arising is what happens when entire traffic flows at a city scale have displays (or indeed other technologies) which potentially give all drivers complete SA? Do benefits at the individual level transfer to the total system level? To explore this issue the individual SA lens needs to be replaced by an exclusively systems SA lens and the final 'Gambon Corner' on the test track taken at full speed.

In a significant departure from the norm Chapter 9 deploys a traffic microsimulation tool to model a range of city-scale street networks. The networks are then populated with several thousand vehicles. These vehicles have varying degrees of awareness of the wider traffic conditions on the network. They can be programmed to have no wider awareness apart from what the simulated driver can see ahead, and to exhibit route choice decisions based entirely on immediate traffic conditions. At the other end of the scale the simulated vehicles can mimic the kind of 'connected' vehicles one would expect to see in a fully realised 'internet of things' or 'smart city' environment: they have complete feedback on the moment to moment traffic conditions on the network and will attempt to optimise their route based on it. They have complete SA. The paradoxes in this case continue to mount. There is a point at which 'more SA' does not give rise to better outcomes in terms of average city-wide journey times, trip lengths and CO_2 emissions. Indeed, for several street pattern types the same overall performance can be achieved with 0% driver SA as it can with 100% driver SA. Pragmatically it is obvious the former is, if nothing else, considerably cheaper than the latter. The good news is it is possible to extract a number of interesting fundamental relationships which can be used to decide what level of collective SA will lead to performance benefits and under what conditions. Relationships of the sort discovered can be used to undertake a new form of user-centred design, one that is applicable to large scale sociotechnical eco-systems of which cars and driving are an exemplar.

The finish line for this research monograph is to have described our 15-year journey into the human factors of vehicle feedback, linked it to the concept of driver situation awareness, and at all times tried to relate the findings from theory to practice. To that end descriptions of pertinent SA methods and a full Goal-Directed Task Analysis for generating SA requirements have been lodged in the Appendix for others to use and modify as they see fit. For automotive engineers and vehicle dynamists we hope to reveal new margins of human performance; new design insights; and ways to access hidden consumer requirements. For human factors readers we hope to have provided a novel perspective on a commonly studied topic; explored and extended the concept of situation awareness; and identified new opportunities to constrain undesirable driver behaviours and enhance others. For any reader who did actually own a 1978 Austin Mini (or indeed Porsche 911 Turbo) we also hope to provide a comprehensive human factors basis for why these cars, and all the other countless billions, felt the way they did and why those feelings – resulting from the interplay of vehicle feedback and driver SA – are important.

2

WHAT'S HAPPENED
TO CAR DESIGN?

2.1 Introduction

2.1.1 Where Does the Feedback Come From and Who Is It For?

The simple reason changes in vehicle design may yield corresponding changes in driver psychology and behaviour is because the vehicle is an intervening variable. It is placed in the control loop between the dynamics of the road environment and the control actions imparted by the driver. The driver's interaction with the environment is indirect. To put it in crude terms, if the driver's interaction with the road environment were literally a direct one, then to stop the car they would place their foot through a hole in the floor and press on the road surface to slow down, rather like the vehicle featured in *The Flintstones* (e.g. Figure 2.1).

Of course, in reality the driver does not press directly on the road surface with their foot in order to slow the car down; they press a pedal. The movement of that pedal is assisted by a vacuum-powered 'servo' unit. This amplifies the force needed to move a piston inside a cylinder. This pressurises hydraulic fluid inside brake lines. This causes pistons in the brake callipers to squeeze brake pads against a cast iron disc. The discs are fitted to the wheel hubs. The retarding force of this action is transferred to rubber tyres. These contact the asphalt road surface, not the driver's shoe. This is now far removed, mechanically, from the actual stresses imposed on the vehicle by this simple action.

The control loop involving all these separate components is also not as simple as it might seem at first. It seems obvious the amount of effort transferred by the driver's foot to the brake pedal should lead to proportional amounts of braking effort, but there is no particular technical reason why a brake pedal could not work in reverse, with lighter foot pressures leading to harder braking. Or that brake pedal

Direct interaction
Kiddimoto wooden balance bike

Driver's feet contact road surface directly, limbs propel vehicle forward.

Indirect interaction
Audi S1 (and virtually every other motorised vehicle)

Driver's feet contact accelerator pedal, throttle potentiometer signals driver requests to electronic control unit, 2.0-litre EA888 turbocharged engine provides rotational torque to 6-speed gearbox and Haldex four-wheel drive system.

Image: Wikimedia Commons

Image: AUDI AG

Figure 2.1 Direct versus indirect interaction with the road environment.

effort versus braking force needs to be continuous. Train and aircraft brakes are not continuous: unlike a car they have a small number of pre-set braking points or levels. Even in a car, while braking might appear linear and proportional, on closer inspection this too is not always the case. There are conditions such as 'brake fade' in which the brakes overheat, and pressing the brake pedal harder will have the reverse effect to that intended (i.e. deteriorating brake performance). In the case of a wheel locking and skidding, and the anti-lock braking system (ABS) becoming active, further increases in braking force are not possible regardless of how hard the brake pedal is pressed. Even within these extreme boundary conditions brake pedal pressure and brake force is rarely linear in a strict sense. Readers may recall vehicles they have personally owned where the brakes seem 'sharp' (small initial brake pedal pressures lead to relatively high brake forces) or 'wooden' (moderate initial pedal pressures lead to comparatively low brake forces). Herein lies the focus of this entire monograph. Within this category of vehicle feedback lie the phenomena frequently referred to as 'vehicle feel', which leads to some powerful human performance effects, both desirable and undesirable.

To explore this issue some further automotive examples can be given. The next is steering feel. Steering feel is analogous to the kind of 'back-feed' pilots receive as they manipulate the primary flight controls in an aircraft (Field and Harris, 1998). The more drastic the control input, and the more extreme the requested aircraft manoeuvre, the harder the control becomes to push or pull or turn. In a completely manual aircraft, where the primary flight controls are physically connected, via rods, linkages and cables, to the flight control surfaces (such as the flaps, ailerons and rudder), this increase in effort stems directly from the amount of rushing air the control surface(s) are pushing against. In much the same way, steering feel in a car arises because the control inceptor (the steering wheel) is physically linked to the system (the collection of steering, suspension and tyre systems) undergoing the stress of converting driver inputs into desired changes in trajectory. In a road vehicle these stresses arise partly from disturbances involving the road surface, from stored energy in the vehicle's tyres, and in particular from a characteristic referred to as aligning torque (Jacobson, 1974).

Aligning torque is an expression of the effort required by the driver to hold the steering wheel in its desired position. Within the normal envelope of vehicle performance the more aligning torque present, or the harder the steering wheel becomes to hold, the more cornering force is being developed by the vehicle's tyres (Jacobson, 1974; Becker et al., 1996). The faster you travel around a corner, the tighter you have to hold the steering wheel. The classic literature in vehicle dynamics describes aligning torque as giving the driver "a measure of the force required to steer the car, i.e. it gives a measure of the 'feel' at the steering wheel" (Joy and Hartley, 1953/4, p. 113). The vehicle itself does not need this feedback, but one only has to imagine the sensation of the steering suddenly going 'light' to realise what an important role it plays for the driver. The irony, however, is that the widespread fitment of power steering often diminishes the effect of aligning torque (e.g. Jacobson, 1974) as do a range of more subtle engineering changes brought about by the almost de facto front-engine/front-wheel drive vehicle platform (e.g. Gillespie and Segel, 1983; Pitts and Wildig, 1978; Bashford, 1978).

Auditory feedback is another good example. It is comprised principally of engine, transmission, tyre and aerodynamic noise (Wu et al., 2003). Numerous studies highlight the role it plays in driving. Despite a noticeable trend to have yet more visually dominant ways of communicating with drivers it has been found they make relatively little use of even existing visual aids, such as the speedometer (Mourant and Rockwell, 1972). A significant part of speed regulation actually relies on auditory feedback. Horswill and McKenna (1999) report that "drivers who received the quieter internal car noise […] chose to drive faster than those who received louder car noises" (p. 983). There is also evidence that lower levels of auditory feedback lead to reduced headway and more risky gap acceptance (Horswill and Coster, 2002). Even more surprising is that completely non-sighted individuals could estimate Time To Collision (TTC) using auditory cues just as well as sighted individuals could using visual cues (Schiff and Oldak, 1990). Again, the vehicle itself does not require auditory feedback. The first patent for an exhaust silencer was filed in 1897 and efforts continue to this day both to minimise engine noise and/or modify its character: not for the car – silencers create back pressure and reduce engine output – but for the driver.

2.1.2 *The Eurofighter Effect*

This disconnect between what the car and the driver needs serves to highlight the problem space we are operating within. In the part of the problem space occupied by the most formal study of vehicle dynamics certain aspects of vehicle function, such as engine noise, could conceivably be ignored completely. In control models that do not have a human element there is simply no reason to include engine noise because no other part of the automotive system needs it. Likewise, a purely technically optimised steering system – one that maximises the vehicle's performance in terms of grip, road holding and agility – would be so heavy to operate as to be unusable, dangerously unstable, or require a solution such as full 'drive-by-wire' in which feedback is artificially created and sent back to the user. This is exactly the situation described by other advanced vehicle types. The Eurofighter Typhoon, for example, "is inherently – and intentionally – aerodynamically unstable [and] requires a complex flight control system to support the pilot" (www.eurofighter.com). The benefits of this 'inherent and intentional instability' are significantly increased manoeuvrability and agility at low speeds, but the aircraft would be unflyable without computer assistance.

The so-called 'Eurofighter Effect' is a term coined by some motoring commentators to explain what they perceive to be a related trend in automotive engineering (e.g. *Autocar*, 2009). Modern vehicles do indeed tend to feel much more responsive and agile than older vehicles. They have objectively higher grip levels, quicker steering and less body roll. The argument runs that with modern electronic stability systems it becomes possible to add a certain amount of 'instability', such as unbalanced front versus rear grip levels to improve 'turn-in' and responsiveness, and rely on the stability system to 'catch' the inevitable loss of rear-end grip if and when it occurs. In support of the 'Eurofighter Effect' is the fact most stability systems cannot now be fully turned off. Not because the vehicle would be undriveable like a Eurofighter, but perhaps it would be a little too unpredictable for the current mass market to accept.

A more subtle version of the 'Eurofighter Effect' might instead be the 'Airbus Effect'. Airbus aircraft are not inherently unstable, far from it; an Airbus' inherent stability, allied to complex computer control, is what helped ensure a relatively safe 'landing' in the river Hudson in January 2009. Despite this inherent stability Airbuses are known for a particular design philosophy, especially when compared to their dominant competitor Boeing. Unlike a manually controlled aircraft in which the pilot's control column and pedals are connected via link rods and cables directly to the flight control surfaces, or in the case of something like a Boeing 737, to hydraulic actuators which move the control surfaces, Airbus controls use a joystick. There are no link rods or cables: pilot commands are sent directly to hydraulic actuators via electrical signals. There is also no feedback through the joystick. It has a relatively lightweight action no matter how fast the aircraft is travelling or how hard it is being manoeuvred. Added to this is that the movement of one joystick (i.e. the captain's) is not mirrored by the other joystick (i.e. the co-pilots). In strict control theoretic terms, why should it? The aircraft does not require the joysticks to become harder to operate the faster the plane is travelling. Neither does the aircraft need both joysticks to be connected so that inputs into one are mirrored in the other. But again, if we venture into other regions of the problem space the human factors issues mount up quickly.

Air France Flight 447 entered a high-altitude stall and crashed due in part to the captain pushing his joystick down while the co-pilot pulled his back, and the aircraft averaging the two inputs (Salmon et al., 2016). Sarter and Woods (1997) report on no fewer than

133 'automation surprises' from a sample of 164 Airbus pilots, the majority stemming from a lack of feedback. Table 2.1 shows the conceptual similarities between this and some modern trends in automotive engineering are quite compelling:

The vehicle as an intervening variable, even one with Airbus levels of authority and autonomy, would be of little concern if drivers (or indeed pilots) were not sensitive to vehicle feedback. The evidence, however, points to the reverse. Hoffman and Joubert (1968) obtained Just Noticeable Difference (JND) data on a number of vehicle handling variables and discovered "a very high differential sensitivity to changes of [vehicle] response time, and reasonably good ability to detect changes of steering ratio and stability factor" (p. 263). Mansfield and Griffin (2000) report that changes in the 'seat of the pants' sensation of just 6% are detectable by normal drivers. Joy and Hartley (1953/4) equate this level of sensitivity to "...the difference in

Table 2.1 Conceptual Similarities between Modern Fly-by-Wire Aircraft and Modern Road Vehicles

AIRBUS A320	INFINITI Q50 (FOR EXAMPLE)
"The Airbus A320 is flown by means of digital controls (fly by wire) – that is, the pilot's input is sent to several flight control computers, which calculate the necessary and allowable adjustments to the flight control surface positions. These computers send their command to hydraulic actuators, which then move the control surfaces" (Sarter and Woods, 1997, p. 557).	The driver's accelerator pedal input is sent to the Engine Control Unit (ECU), which calculates the necessary and allowable adjustments to the fuel/air delivery system. The ECU sends its commands to electronic actuators, which then move the control surface (the throttle plate/fuel injector solenoids/turbo wastegate control etc.).
"This design allows for the introduction of new functions such as envelope protection, which prevents the pilot from exceeding certain limits of the flight envelope. In other words, a fly-by-wire design assigns more authority and autonomy to the automation" (Sarter and Woods, 1997, p. 557).	This design allows for the introduction of new functions such as stability control, anti-lock brakes and speed and rev limiters, which prevent the driver from exceeding certain limits of the drive envelope. In other words, a drive-by-wire design assigns more authority and autonomy to the automation.
"The distinct design of flight controls on this aircraft (i.e. the uncoupled sidesticks and the non-moving throttles) has raised concerns about potential negative effects of removing peripheral visual, tactile, and auditory cues, as these may help pilots monitor automated system activity and maintain [...] awareness" (Sarter and Woods, 1997, p. 558).	The distinct design of driving controls on this vehicle (i.e. the steer-by-wire steering and paddle gear shifters) has raised concerns about potential negative effects of removing peripheral visual, tactile and auditory cues, as these may help drivers monitor system activity and maintain awareness.

feel of a medium-size saloon car with and without a fairly heavy passenger in the rear seat" (Joy and Hartley, p. 119).

Herein lies the paradox. The high level of driver sensitivity to feedback stands in contrast to the increasing power, authority and autonomy of current vehicle systems, let alone those in future which are likely to emulate Airbuses and Eurofighters even more closely. In other words, the panoply of advanced vehicle systems, from automation to new drivetrains, will exert an influence far greater than the sensitivity threshold of drivers to be able to detect it (e.g. Walker et al., 2001; Bedinger et al., 2016). Moreover, these changes in vehicle design are often in the direction of isolating the driver even more from the subtle moment to moment state of their vehicle in its environment (Loasby, 1995; Norman, 1990; Zuboff, 1988). Why should we be interested? Because if the Eurofighter/Airbus Effect is true then we already know – and have known for over 20 years – that "the distinct design of flight controls on [fly-by-wire aircraft] has raised concerns about potential negative effects of removing peripheral visual, tactile, and auditory cues, as these may help pilots monitor automated system activity and maintain situation awareness" (p. 558). In order to begin exploring the issue of vehicle feedback and driver situation awareness then a good place to start would be a motoring equivalent of a study comparing Airbuses and Boeings. Such a study is described below.

2.2 What Was Done

2.2.1 Design

This experiment was exploratory and designed as an initial investigation into the issues at hand. It is based on a naturalistic on-road driving paradigm in which individuals used their own vehicles to drive around a defined route on public roads. As they did so, drivers provided a 'concurrent verbal commentary'. Key constructs relating to individual and systemic SA were extracted from the transcript of their commentary using a theme-based content analysis. The verbal protocol was dependent upon one independent variable (vehicle type) with two levels: high feedback car (median design age 1983 and fewer driver assistance systems) or low feedback car (median design age 1998 with more driver assistance systems). The two groups of vehicle – high and low feedback – were separated by a mean of 15 years in vehicle design evolution, although in individual cases this difference extended to more than 30 years. The study also took place at a unique time because sufficient numbers of vehicles without power steering (an almost universal feature at the current time of writing) were available in the vehicle fleet to be sampled. A self-report Driving Style Questionnaire (DSQ; West et al., 1992) was administered to drivers in order to control for potential biases in car choice (do faster drivers, for example, choose newer cars?). Measurements of average speed and time were also taken in order to control for potential biases in how fast the course was driven and the 'objective' style in which it was done so. Finally, demographic variables were used to, again, control for possible biases in the type of vehicle owned and whether this systematically varied depending on, for example, participant age. All experimental trials took place at times designed to avoid peak traffic conditions for the area. These times were defined through extensive pilot trials and route testing runs. The experimental runs themselves all took place in dry, clear weather.

2.2.2 Participants

For this opening study, 10 car drivers took part using their own vehicles, a deliberate step to ensure familiarity with the characteristics of the vehicle. Over-sampling was used in order to pre-select a sample well-matched in age and experience, but with differences in the characteristics of the vehicle being driven that were sufficiently large and differentiating. All participants held a valid UK driving licence with no major endorsements and reported they drove approximately 12,000 miles per year, which is the UK average. The participants' ages fell within the range 20 to 50 years old. Mean driving experience was 14 years (minimum 3 years, maximum 44 years) with all the drivers having been exposed to many hundreds of hours driving. In this case an all-male sample was used. This was a purely pragmatic step based on those volunteering and also to ensure within the context of this small sample that gender differences could not provide an unwanted confounding variable. There is no reason at this stage to suspect significant differences according to gender (e.g. Walker et al., 2001). It is also important to state that having discovered the challenges of recruiting gender balanced samples for studies of this type extra efforts were made in subsequent studies to correct for it. All drivers were compensated for their participation and they did so with informed consent and with ethical approval granted by the host institution.

2.2.3 Materials

2.2.3.1 Vehicles The study took place at a point in time, and with a population, whereby the vehicle fleet still contained a useful proportion of older vehicles. These were lacking certain important vehicle technologies then becoming commonplace, and now almost universal. It would be difficult to perform a similar study today. The vehicles used in the experiment are detailed in Table 2.2.

Without wishing to 'overuse' (or indeed abuse) the Boeing versus Airbus analogy, it remains possible to draw out some relevant parallels in our vehicle sample. Taking the Boeing 737 and Airbus A320 as a rough guide we can observe approximately 20 years in their respective design ages. The 737 was a 1960s design whereas the Airbus was launched in 1984. In a similar manner, the vehicles used in this study include a group with a mean 'design age' of 1983 and another collection with a mean of 1998, a difference of 15 years. The 'design age' in this case is the year the vehicle type and mark first became available on the UK domestic market as an official import (Robson, 1997). Analysis of vehicle specifications for the two vehicle groups, based on the median, shows the modern vehicles are 19.36% heavier but 34.71% more powerful, with a 17.5% faster top speed. This reflects wider trends in the overall vehicle fleet well (e.g. Martin et al., 2017). Crucially, and to work

Table 2.2 Sample Vehicles

'BOEINGS'	AGE	'AIRBUSES'	AGE	VEHICLE TYPE
Peugeot 309 GLD	1988	Volkswagen Golf TDi	1998	Diesel Hatchback
Renault 18 GTX	1978	Mitsubishi Space Wagon	1992	Estate/People Carrier
BMW 325i	1983	Holden HSV	2000	Sports Saloon
Morgan 4/4	1968	Maserati 3200	2001	Sports Convertible
Toyota Tercel	1985	Volkswagen Golf CL	1992	Hatchback
Median Design Age	1983		1998	

the Airbus/Boeing analogy further, none of the older cars are fitted with power steering. They are 'manually controlled' vehicles with a direct mechanical linkage between the steering wheel and front wheels. The newer cars, conceptually 'the Airbuses', all have power steering. Whilst still having a direct mechanical link it is hydraulically assisted. As previously discussed, this alters the 'back-feed' to the control in a conceptually similar way to the lack of backfeed in an Airbus. The modern cars also have many of the more advanced forms of 'envelope protection systems' including ABS (e.g. Volkswagen Golf), traction control (e.g. Maserati) and rev limiting (all). Like their Airbus counterpart they also benefit from 15 years of predominantly market-driven improvements in vehicle comfort and refinement. The Morgan 4/4 is a particular case in point, being an open-top sports car with a wooden chassis and a design that in certain critical respects (e.g. the use of 'sliding pillar' suspension) dates from the 1930s. If it were a Boeing it would be closer to a Model 40 bi-plane rather than a 737, hence the importance of not stretching the analogy too far.

2.2.3.2 Apparatus Audio and video data was captured from the driving scenario using a miniature video camera and laptop computer. The view provided is illustrated in Figure 2.2.

2.2.3.3 Other Materials Standardised instructions were provided to each participant. These instructions prepared the participant for the route they would be driving, the role of the experimenter (who would only provide directional instructions during the drive), and guidance on the desired form and content of the verbal commentary ("think aloud about the things around you relevant to the current driving task" and "keep talking, even if what you are saying does not appear to make much sense"). Explicit written instructions were also devised on the criteria to be used for completing the thematic content analysis. This was for use by the experimenter(s) during the analysis phase in order to benefit inter- and intra-rater reliability. One experimenter accompanied each driver on the route, and three analysts took part in the content analysis reliability checking process. Driving style was assessed via a self-report driving style questionnaire (DSQ; West et al., 1992).

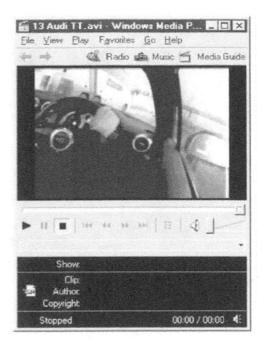

Figure 2.2 Forward view from the video apparatus.

2.2.3.4 The Test Route The on-road route was contained in the West London area of Surrey and Berkshire in the UK. It was 14 miles in length not including an initial three-mile stretch used to warm up participants. The route devised for this study represented a compromise between ensuring drivers were exposed to a wide range of road types, and restricting the concurrent verbal commentary to approximately 30 minutes in length to avoid fatigue and response bias. The route comprised one motorway section, seven stretches of A or B classification roads (single and dual carriageway trunk roads), two stretches of unclassified roads (minor country roads), three stretches of urban roads (main roads through busy built-up areas), one residential section and fifteen junctions. Figure 2.3 presents a diagram of the route and its primary features.

2.2.4 Procedure

Formal ethical consent was obtained from all participants before the study commenced. Particular emphasis was given to the participants' responsibility to ensure their safety and that of other road users.

Figure 2.3 Map of the 14-mile on-road course.

Participants then completed the DSQ followed by a comprehensive experimental briefing. The concurrent verbal protocol consisted of the driver providing a 'running commentary' about the information they were taking from the driving scenario and how they were putting it to use. An instruction sheet on how to perform a concurrent verbal protocol was read by the participant, and the experimenter provided verbal examples of the desired form and content. The three-mile approach to the start of the test route enabled participants to practise and to be advised on how to perform a suitable concurrent verbal protocol. During the data-collection phase the experimenter remained silent apart from offering route guidance and monitoring the audio/video capture process. The experiment was complete upon return to the start point.

2.3 What Was Found

2.3.1 Treatment of the Data

A two-stage approach to the statistical exploration of the data was taken. Firstly, a conventional approach based on inferential statistics was used. In this case non-parametric techniques were deployed as they do not rely on assumptions the data is normally distributed.

They lend themselves well to smaller sample sizes such as this. Non-parametric techniques are also based on converting raw numerical data into ranks and therefore utilise the median as opposed to mean as a measure of central tendency. This again helps to mitigate the effects of a small sample. Non-parametric tests also represent a more cautious test of inference as they are not generally as powerful as parametric tests. Combined, this represents a conservative approach appropriate to exploratory work of this kind.

All approaches to inferential statistics are dependent on sample size. Typically, as the sample size increases so does the probability of obtaining a significant result, and vice versa. It also provides little indication of the magnitude of difference brought about by the intervention of vehicle type. The solution, and the second of two stages, is to compliment inferential statistics with an alternative statistical approach based on effect sizes. An effect size (ES) index based on a simple correlational analysis is performed on the independent variable (vehicle type) and the results of the various measures of SA (the dependent variables; Coe, 2002). The higher the correlation the more that vehicle feedback can be said to have an impact on driver SA. The procedure is outlined in Howitt and Crame (2016). Overall, the two-step approach was designed to offer a pragmatic and rational exploration of the data, and a basis upon which further hypotheses and larger scale studies could be developed.

2.3.2 Analysis of Control Measures

Table 2.3 and Table 2.4 present the descriptive, inferential and effect size correlation analyses of the median speed taken to complete the experimental road course (and by implication the time as well) and the outcome of the DSQ. The statistical method used to detect the probability of significant differences between the two vehicle groups on these two dimensions was a Mann–Whitney U test (notated as U).

Statistical power is not sufficiently high in either case of median speed and DSQ score to assume there is literally zero difference between the drivers of the two vehicle types. The effect size correlation statistic (notation R_{bis}) suggests instead that approximately 4% of the variance in DSQ scores is explained by vehicle type (notated R^2) albeit not statistically significant at the 1%, 5% or 10% level. In other

Table 2.3   Summary of Analysis for the Control Measures

	NEW CAR	OLD CAR
Mean Overall DSQ Score	3.13	3.07
Time to Complete Route	26 min 01 sec	28 min 01 sec
Average Speed	31.60 MPH	29.47 MPH

Table 2.4   Inferential Statistics Comparing Vehicle Age with the Control Measures

	TEST STATISTIC (U)	PROBABILITY	EFFECT SIZE (R_{BIS})	VARIANCE EXPLAINED (R^2)
Mean DSQ Score	12.00	p = ns[a]	0.19	0.04 (4%)
Route Time/Speed	9.0	p = ns[a]	0.41	0.17 (17%)

[a] = not significant at the 1%, 5% or 10% level

words, there is a very small (and non-significant) relationship between a person's driving style and the vehicle they own/drive, but at 4% it is small enough to be of little practical importance. On the other hand approximately 17% of the variance in median speed is explained by vehicle type, despite the lack of statistical significance at either the 1%, 5% or 10% level. The newer cars are being driven slightly faster. Although there is clearly an effect of vehicle type on speed driven, albeit non-significant, it is important to note the variance is only around 2 mph and can once again be considered of minimal practical impact. Taking the most generous of interpretations and viewing the 'direction' of the effect (i.e. reduced auditory feedback inside newer, quieter cars) it seems to accord with the literature. Reduced auditory feedback gives rise to higher vehicle speed. Again, this is not a particularly safe assertion to make in this case and it seems more reasonable, in the context of this exploratory study, to conclude that the two groups are well matched on these control measures. Drivers of older and newer vehicles are not inherently or meaningfully different to each other. Any significant effects on driver SA of vehicle design can be exposed should such differences actually exist.

2.3.3 Analysis of Concurrent Verbal Protocol Data

The verbalisations captured by the audio/video equipment were transcribed verbatim against a two-second incremental timeline. A content analysis was then undertaken. It centred on the extraction of content words from function words. Function words are bracketed in the following extract as an example: "[passing] [driveway] on [left], [50 mph], thinking about [changing] [up] a [gear] to [5th]...." Function words describe items of information being referred to by drivers. This helps to reveal the nature of their SA. The encoding scheme was anchored to specific SA categories and is 'exhaustive'. When a function word meets the definitions described in the encoding instructions it enters into the analysis, irrespective of whether it is already represented in other categories.

After completing the encoding process for all participants the reliability of it was then established. Two independent raters encoded previously encoded analyses in a blind condition using the same categorisation instructions the original rater employed. Across the seven individual encoding categories inter-rater reliability (IRR) was established at R_{ho} = 0.7 (n = 756) for rater number one and R_{ho} = 0.9 (n = 968) for rater number two. Both of these correlation values were statistically significant ($p < 0.05$ for rater one and $p < 0.01$ for rater two). Intra-rater reliability was also examined to check for any 'drifting' in encoding performance over time. This analysis posted a correlation of R_{ho} = 0.95; n = 756; $p < 0.01$, suggesting that encoding performance over time was, indeed, remaining stable. With these potentially confounding variables adequately controlled it is now possible for various features of driver SA to become exposed.

2.3.3.1 Individual Driver SA The first 'SA lens' to be slid across the analytical frame is that of individual SA. This analytical lens describes early models of SA well. Heavily psychology-based, such models attempted to explain the processes underpinning the awareness held in the minds of individual people (Stanton et al., 2010; Endsley, 2015; Figure 2.4).

If SA is viewed simply as the sum of knowledge relating to specific topics within the system (e.g. Baber, 2004) then it seems drivers of the older-generation 'Boeing' cars are extracting more knowledge objects from the scenario and/or using the same ones more often than drivers of the more modern 'Airbus' vehicles. A total of 6823 encoding points were derived from the transcripts of the concurrent verbal protocols. Drivers of the manual, older generation high-feedback cars provided a total of 3924 data points compared to 2899 for drivers of the modern low-feedback cars. This difference is statistically significant (U(N1 = 5,

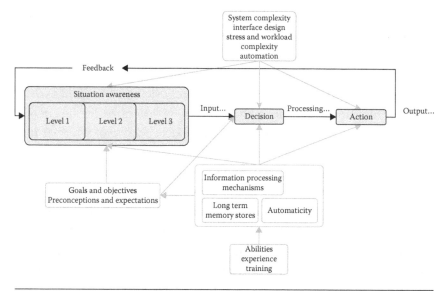

Figure 2.4   Simplified form of the Endsley (1995) three-level model of SA.

N2 = 5) 3, $p < 0.05$) and a strong effect size was detected, $R_{bis} = 0.6$. An R^2 value of 0.36 means approximately 36% of the variance is being explained by vehicle feedback. Subsequent chapters reveal some important subtleties in this argument – 'more' does not necessarily mean 'better' – but for present purposes it can be allowed to remain. It illustrates clearly the perceptual impacts of vehicle feedback.

Theories of individual SA go into detail about the psychological processes involved (e.g. mental models, perception, comprehension) and the nature of SA itself (e.g. the person's SA comprises knowledge about x, y and z). Endsley's three-level model (1995a) has undoubtedly received the most attention. This is an information processing–based model which describes SA as an internally held cognitive product comprising three levels: perception (Level 1), comprehension (Level 2) and projection (Level 3), all of which feed into decision-making and action execution. Level 1 SA involves perceiving the status, attributes and dynamics of task-related elements in the surrounding environment (Endsley, 1995). According to the model a range of factors influence the data perceived, including the task being performed and the individual's goals, experience and expectations, as well as systemic factors such as interface design, level of complexity, automation and feedback. To achieve Level 2 SA, the individual interprets the Level 1 data and comprehends its relevance to their task and goals. Level 3 SA involves forecasting future system states using a combination of Level 1 and 2 SA-related knowledge, and experience in the form of mental models. By these means drivers can forecast likely future states and use this to take action. Endsley's model foregrounds cognitive models of information processing, and evidence of this can be seen in the loose parallels that can be drawn between more basic Input-Processing-Output models of cognition (Stanton et al., 2001). Endsley's model is of course more complex and nuanced than this, but the hereditary line is apparent and entirely consistent with the zeitgeist of cognitive psychology evident at the theory's inception in the 1980s. Framed in this way it is possible to turn to the verbal transcripts provided by the drivers in this study and use Levels 1 (Perception), 2 (Comprehension) and 3 (Projection) as encoding themes.

Figure 2.5 presents the outcomes of this analysis, showing the results partitioned into Levels 1, 2 and 3 consistent with Endsley's model of SA. Table 2.5 presents the results of the corresponding statistical analysis. It can be seen that Level 1 SA differs significantly between the drivers of the two vehicle groups (U(N1 = 5, N2 = 5) 63.5, $p < 0.1$) and in a direction which suggests 'more' SA for drivers of higher feedback cars. Once again, the effect size correlation is suggestive of a large effect (Table 2.5). Based on Figure 2.5 it is not fully clear whether the

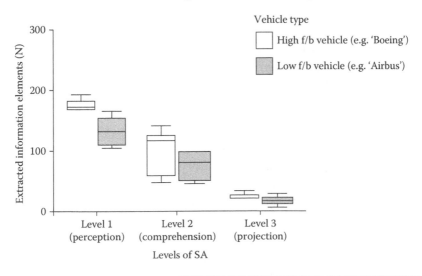

Figure 2.5 Boxplot illustrating distribution and number of extracted knowledge objects for each level of Endsley's (1995) three-stage model of SA according to vehicle age.

Table 2.5   Summary of Analysis on Individual SA According to Vehicle Type

	LEVEL 1 SA PERCEPTION	LEVEL 2 SA COMPREHENSION	LEVEL 3 SA PROJECTION
Non-Parametric Statistic (U)	3.5	10	9
P-Value	$p = 0.06$	$p = ns^a$	$p = ns^a$
Approximate Effect Size (R_{bis})	0.54	0.23	0.29
Variance Explained (R^2)	0.29 (29%)	0.05 (5%)	0.08 (8%)

[a] = not significant at the 1%, 5% or 10% level

relatively insulated drivers of the newer cars are significantly disadvantaged in terms of Level 1 SA compared to drivers of the older 'manual' cars, or if drivers of newer (or indeed any) cars are able to comprehend the situation and project onto future states using less frequent occurrences of Level 1 SA. There are no similar differences between the vehicle types for Level 2 and 3 SA. The possibility exists that drivers can achieve these comparative amounts of Levels 2 and 3 awareness on significantly different amounts of Level 1 SA. In other words, drivers can compensate. The study reported in Chapter 8 sheds further light on this issue but suffice to say that even 'if' this is the case, the limits of this compensation are currently unknown. How much vehicle feedback can be removed before driver performance (and the driving experience) materially suffers? At present we do not know.

2.3.3.2 Systemic Driver SA

The individual SA lens can now be replaced with a systems SA lens and a different set of results projected. Systems thinking in general presents some distinct challenges for individualistic views of SA. One of the main ones is simply that SA becomes a phenomenon which emerges from the interaction of multiple 'agents' in a system, and this emergence cannot be viably studied using reductionist (i.e. individualistic) approaches. SA shifts from something that resides exclusively in the minds of individuals to instead a network of information on which different components of the system have distinct views and ownership.

Transactive SA describes the notion that SA is acquired and maintained through transactions in awareness arising from communications and sharing of information. A transaction represents an exchange of SA between one agent and another, where agent refers to humans, artefacts and environments. As agents in the system receive information it is integrated with other information and acted on, and then passed onto other agents. These features find their clearest expression in the Distributed SA (DSA) concept (Stanton et al., 2009; Salmon et al., 2009) which is used in Chapter 7. For now we can highlight where, in the joint cognitive system of driver, vehicle and road, SA resides and is coming from. To this end drivers' verbalisations were categorised into the following SA agent categories: Own Behaviour (Behaviour), Behaviour of the Car (Vehicle), Road Environment (Road), and Other Traffic (Traffic). Figure 2.6 shows the results as a profile, illustrating the proportion of total information objects which fall into these four categories. Differences in the profiles between the two vehicle types illustrate the structure and type of information, and its source, differs. This implies the type of SA held by the respective drivers will also differ, as indeed Chapters 6 and 7 show.

Statistically significant differences across the SA agent profile were detected for drivers of the two vehicle types using a Friedman test (Chi Square = 8.76, df = 3, $p < 0.05$ and Chi Square = 14.04, df = 3, $p < 0.01$ respectively). Multiple statistical comparisons between the four agent categories (using a procedure outlined in Siegel and Castellan, 1988) reveal a pattern of significant differences that again differs according to vehicle type (Table 2.6).

Drivers of the high-feedback cars are reporting significantly more knowledge objects in all of the agent categories than drivers of the low-feedback cars, as illustrated in Figure 2.6. Statistically significant differences were detected between high and low feedback car drivers in the behaviour category (U(N1 = 5, N2 = 5) 4, $p < 0.1$), the vehicle category (U(N1 = 5, N2 = 5) 3, $p < 0.05$) and the other traffic category (U(N1 = 5, N2 = 5) 3, $p < 0.1$), but not the road environment category. Drivers of high-feedback vehicles could be said to have greater SA than drivers of low-feedback vehicles in that they have – at a simplistic level – greater awareness. Large effect sizes were also in evidence. The most interesting finding concerns the active information elements in the vehicle category. The high-feedback vehicles are providing more information than the low-feedback vehicles – more driver feedback – and are contributing more to driver SA. The statistically significant results for behaviour, vehicle and other traffic categories are presented in Table 2.7.

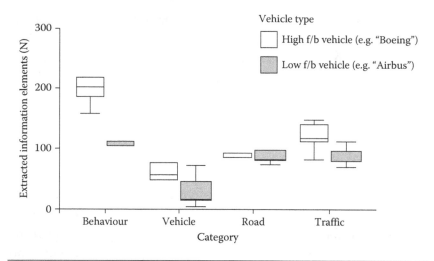

Figure 2.6 Boxplots showing the number and category of extracted information elements for the low f/b 'Airbus' cars and the high f/b 'Boeing' cars across different parts of the driving system.

Table 2.6 Summary of Multiple Comparisons across the Profile of Systemic SA for Drivers of New and Old Cars

COMPARISONS	MEAN RANK LOW FEEDBACK	MEAN RANK HIGH FEEDBACK
Behaviour-Vehicle	2.4 ($p < 0.01$)	3 ($p < 0.01$)
Behaviour-Road	1.0 ($p =$ ns[a])	1.8 ($p < 0.01$)
Behaviour-Traffic	1.0 ($p =$ ns[a])	1.2 ($p =$ ns[a])
Vehicle-Road	1.4 ($p < 0.1$)	1.2 ($p =$ ns[a])
Vehicle-Traffic	1.4 ($p < 0.1$)	1.8 ($p < 0.01$)
Road-Traffic	0.0 ($p =$ ns[a])	0.6 ($p =$ ns[a])

[a] = not significant at the 1%, 5% or 10% level

Table 2.7 Summary of Analysis on the Quantity of SA According to Vehicle Type

	BEHAVIOUR	VEHICLE	TRAFFIC
Non-Parametric Statistic (U)	4	3	4
Probability Value	0.08	0.05	0.08
Effect Size (R_{bis})	0.60	0.55	0.56
Variance Explained (R^2)	0.36 (36%)	0.30 (30%)	0.31 (31%)

The key differences in the content of SA according to vehicle type are as follows. Drivers of high-feedback vehicles report more information elements about their own behaviour than they do about the road environment (there is no similar statistical difference for drivers of low feedback vehicles). Drivers of the low-feedback vehicles talk significantly less about their vehicle than they do about the road (there is no similar difference for drivers of high-feedback cars).

2.3.4 Do the Findings Matter?

Figure 2.7 presents an effect size summary. Effect is related to the importance of the finding. The larger the effect, the greater the potential 'real-world' importance. This is especially useful given the exploratory nature of this study. If large effects are manifest despite a small sample then areas of future interest have been revealed. What the findings show is that vehicle design and consequent changes in driver feedback explain close to 0% of the variance in driving style, but possibly 17% of the variance (albeit only 2 mph) in median speed around the on-road course. This might be expected given the extra performance available from the modern vehicles, and it is also consistent with the literature on the varied effects of reduced vehicle feedback on driving performance, in particular that quieter cars have been shown to be driven more quickly. Much larger effect sizes are in evidence for the independent variables of individual and systemic SA. These sample

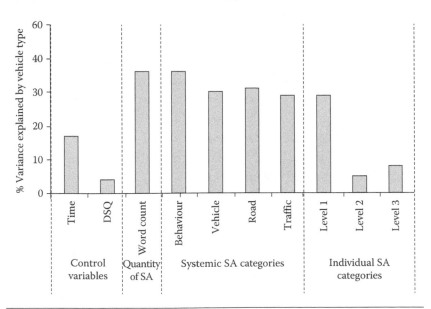

Figure 2.7 Effect size summary showing the % variance explained in each of the control and dependent variables based on a comparison of vehicle type.

effects are suggestive of the direction and strength of population effects when vehicle design is manipulated and driver SA measured. Figure 2.7 therefore points to areas where this initial exploration will build into more comprehensive insights as the book progresses, and where, ultimately, vehicle design has (and could in future have) the largest impact on driver SA.

2.4 Summary

This monograph began by presenting the dramatic extent of change in vehicle performance and design, and asking whether it would be a surprise if such changes did not have an impact on the driver. Clearly the answer is they do. We also mentioned a number of well-worn anecdotes and asked whether they may in fact be hinting at something more profound. The answer is yes. The amount of information extracted from the driving context, and in particular the vehicle, is certainly reduced for drivers of low-feedback cars. The anecdotes, therefore, do indeed seem to point towards deeper, more fundamental issues which warrant further exploration.

While the current findings are exploratory in nature, and designed to provide an entry point into the issues at hand, they are consistent with what has gone before. They find support in the work of Hoffman and Joubert (1968) by showing that a 15-year gap in vehicle design is clearly within the sensitivity threshold of normal drivers. They also support the notion that all manner of 'hard engineering' vehicle traits do indeed have an impact on driver behaviour and perceptions (e.g. Joy and Hartley, 1953/4; Pitts and Wildig, 1978; Bashford, 1978; Gillespie and Segel, 1983; Barthorpe, 2000; Horswill and McKenna, 1999; Bosworth et al., 1996; Joy and Hartley, 1953/4; Jacobson, 1974; Nakayama and Suda, 1994 and so on).

Taking a step further back is seems clear the feedback characteristics of almost all cars – fundamentally – are well-aligned to user-centred principles. It does not seem excessive to say that car design has achieved the kind of 'appropriate feedback' described by Norman (1990) in his seminal 'Ironies of Automation' paper. The vehicle, via its innate feedback characteristics, is providing the kind of timely 'informal chatter' (in a psychological sense) which keeps the driver attentive and informed of their situation. The paradox for vehicle design is that having achieved this desirable state, future vehicles have the potential to remove it completely. This could remove the very characteristic of driving which control theory and psychology would argue helps to ensure that, despite all the inherent complexity involved in driving a car, it can still be carried out with relative ease and safety by most drivers (Milliken and Dell'Amico, 1968). Such a situation, which is likely and forecasted (e.g. Walker, Stanton and Young, 2001; Bedinger et al., 2016) would closely emulate an Airbus-versus-Boeing situation. This situation has been studied for over 20 years in aviation and we are well aware of the potential performance problems which can arise. There seems to be much to learn, perhaps even re-learn, about the current interaction drivers have with vehicles before moving ahead to consider how this can be improved with technology. Certainly the call to explore more of the vehicle dynamics problem space seems justified.

3

FEEDBACK PROPERTIES OF VEHICLE CONTROLS

3.1 Introduction

Are cars one of the most complex mechanical devices available for consumers to use and buy? With well over 150,000 component parts, quite possibly. With vehicle design changing so much, and some emerging insights into what these changes might mean for the driver, perhaps now is a good time to revisit the not-so-humble car and ask, from a Human Factors point of view, what sort of device it actually is. The ubiquity of the humble car tends to conceal the feat of high-technology mass production it represents. Although the underlying mechanical systems are largely unseen they are impressive in their capabilities. Most impressive is that all the driver has to do in order to harness this immense energy and performance is rotate a wheel with their hands and press pedals with their feet.

3.1.1 Manual Control

The vehicle's fundamental systems are controlled by the driver using various control inceptors. Vehicle steering (fundamental system) is controlled by the steering wheel (inceptor); engine speed and power output (fundamental system) are controlled by the accelerator pedal (inceptor); and vehicle brakes (system) are controlled by the brake pedal (inceptor). Broadly speaking, what all of these controls have in common is the position of the control inceptor is more or less directly related to the rate of system output (Milliken and Dell'Amico, 1968). Steering wheel angle is proportional to the rate of turning, accelerator position is proportional to the rate of engine power delivery, and brake pedal position is proportional to braking force. Within the scope of normal driving this proportional and predictable relationship holds broadly true.

Ergonomically speaking, driving would be regarded as a manual control task. Manual control, in the form of moving limbs and hands to operate control inceptors, represents the output of human information processing. Tracking refers to when this output has to be continuous in order to keep the system within certain parameters. Milliken and Dell'Amico (1968) point out that 'rate' control of the sort found in cars is intuitive. There are a number of strategies which can be employed to aid human performance in tracking tasks, and whether by design or evolution, vehicle feedback provides many of these aids. They will be dealt with in more detail in the sections that follow. For the time being, rate control and vehicle feedback provide an explanation for how an ostensibly complex mechanical system such as a car can be controlled so easily by such a diverse population of drivers. In fact, it has been suggested that riding a bicycle presents equal if not more control difficulties (Milliken and Dell'Amico).

3.1.2 *The Tracking Loop*

The driver is part of a tracking loop involving a command input, given the notation $ic(t)$. The command input is represented by the performance of specific driving tasks, and is issued in response to some form of error which needs to be neutralised, given the notation $e(t)$. Errors, in this sense, are cues from the environment telling the driver they are deviating from a planned trajectory unless some action is taken. The force applied to the relevant control (such as the steering wheel) is expressed as $f(t)$, and the movement of the control in relation to the force $f(t)$ is expressed as $u(t)$. The relationship between $f(t)$ and $u(t)$ is termed *control dynamics* (Wickens, 1992).

Control dynamics express the relationship between the force applied and the movement of the control, or the reaction of the control to an input from the driver (Wickens). Obviously this relationship will be influenced by feedback through the control from the environment, and this can be notated as $id(t)$. For example, the driver applies a rotational torque to the steering wheel which causes it to turn. This action overcomes any resistance within the vehicle's steering and front suspension system, and works against any feedback from the road. The response of the car to the control input ($f(t)$) is termed $o(t)$, and this is termed *system dynamics* (Wickens, 1992). System dynamics is the relationship between control position and the output of the system. In the present example this output relates the position of the steering wheel to the lateral acceleration of the vehicle within the external world.

The relationship between any system and its control dynamics can be expressed as a transfer function. A transfer function presents the mathematical relationship between the input from the driver and the output provided by the car. The relationship between control and system dynamics – the driver's inputs and the vehicle's outputs – can be analysed according to a collection of these transfer functions, and they reveal what sort of user-centred entity we are dealing with.

3.2 Powertrains

3.2.1 Engine Basics

The engine sets the vehicle in motion, and in doing so initiates all the forces other passive vehicle systems (like suspension and brakes) have to contend with. Engines of whatever form are simply energy converters (Nunney, 1998). At the time of writing the internal combustion engine is still by far the dominant 'prime mover' fitted to passenger vehicles. These extract the chemical energy stored within petrol or diesel by igniting a mixture of fuel and air using a four-stroke cycle of induction (suck), compression (squeeze), ignition (bang) and exhaust (blow) (Figure 3.1).

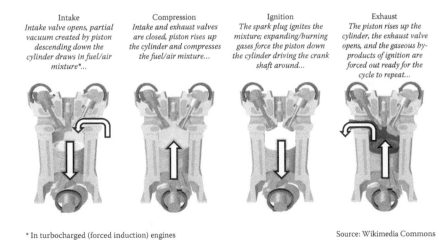

Intake	Compression	Ignition	Exhaust
Intake valve opens, partial vacuum created by piston descending down the cylinder draws in fuel/air mixture*...	*Intake and exhaust valves are closed, piston rises up the cylinder and compresses the fuel/air mixture...*	*The spark plug ignites the mixture; expanding/burning gases force the piston down the cylinder driving the crank shaft around...*	*The piston rises up the cylinder, the exhaust valve opens, and the gaseous by-products of ignition are forced out ready for the cycle to repeat...*

* In turbocharged (forced induction) engines the fuel/air mixture enters at greater than atmospheric pressures.

Source: Wikimedia Commons

Figure 3.1   Four-stroke cycle of induction, compression, ignition and exhaust for petrol engines.

Different methods of extracting chemical energy are needed depending on the fuel used. Petrol needs to be mixed in a precise, relatively constant 'stoichiometric' ratio of air and fuel, typically 14.7 parts air to one part petrol (Schwaller, 1993). The air/fuel mixture needs to be compressed and ignited with a spark, which then leads to an extremely rapid burning process (as distinct from the common misnomer of an 'explosion'). The burning process causes the petrol and air to be converted into various exhaust gases. It is the heat and rapid expansion of these exhaust gases which pushes a piston down a cylinder, causing a crankshaft to rotate, and a twisting force to be made available to ultimately turn the vehicle's wheels. In petrol engines the accelerator pedal is connected to a throttle which restricts the amount of air–fuel mixture admitted into the combustion chambers. In diesel engines there is no throttle (or restriction) and engine power output is varied by how much fuel is injected. By these means the driver can control the engine's power output and speed, and in turn the vehicle's speed and acceleration.

Engine power is measured in different ways. Strictly speaking engine power is described by engine torque. Torque is defined as 'twisting force' (Schwaller, 1993; Nunney, 1998) and can be measured directly from the rotating engine crankshaft. Torque is typically expressed as the energy needed to move a certain number of pounds one foot (lb ft.) or Newtons over metres (Nm). For example, the maximum twisting force the engine in a Mini One is able to deliver is 133 lb ft. (or 180 Nm), which is the same as applying a 133-lb weight to a foot-long lever (or applying a force of 180 Newtons to a metre-long lever).

The Mini One also produces 102 brake horsepower between 4000 and 6000 revolutions per minute. Brake horsepower (BHP) is a commonly quoted measurement, and the formula for horsepower is torque multiplied by engine speed in revolutions per minute, divided by 5252. The 'brake' refers to the measurement method. The engine is worked against some form of pre-defined braking force, and the 'unknown' amount of power it develops is measured against this 'known' braking effect. Horsepower did originally refer to the work performed by a horse. The Victorian steam engineer James Watt (from which the lead author's institution, Heriot-Watt University, derives half of its name) determined that a 'standard' horse could lift an average of 550 pounds in weight one foot in one second, or 33,000 pounds one foot in one minute (Schwaller, 1993). A number of modern variations on the 'standard horse' have been derived including the Society of Automotive Engineers (SAE) horsepower and the Deutsche Industrie Norm (DIN) horsepower. The former theoretical horse is slightly smaller than the latter; indeed, early automotive engineers were perturbed to discover that some real-life horses produced as much as 10 horsepower.

The official measuring unit in Europe and beyond is the Watt or kW for most vehicle engines. One metric horsepower is equal to 735.5 watts, making the Mini One possess a 75-kW engine. Ignoring any conversion losses, and purely to give a feel for what 75 kW equates to in 'real-life' terms, it is sufficient to provide the electrical needs of approximately four homes or equivalent to the power of approximately 140 e-bikes. A not-inconsiderable amount of power, therefore, is needed for even quite modest vehicle performance.

An important characteristic of all petrol or diesel reciprocating piston engines is they need to be running at a given crankshaft speed in order for their useful power output to be derived. Compare this to electric motors. These can exert their maximum force whilst completely stationary. Different engine technologies give rise to different power delivery characteristics, described in a power curve of engine speed versus power or torque output (Figure 3.2). This explains why a petrol or diesel-powered car will stall if the engine speed is too low for the demands being placed on it (i.e. not enough power) or cause a loss of traction if the engine speed is too high for the conditions (i.e. too much

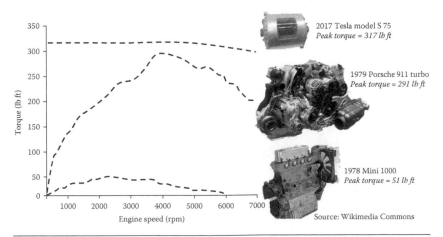

Figure 3.2   Torque curves for 1978 Mini, Porsche 911 Turbo and Tesla Model S 75.

power). Between these two extremes is a relationship between a driver's control inputs and the engine's outputs described by its power curve.

3.2.2 Vehicle Speed Control as a Tracking/Manual Control Task

Controlling vehicle speed, and by implication engine power output, can be described as a tracking task. In driving the task is self-paced and involves imparting correct control operations at the correct time (Sanders and McCormick, 1993). Operating the vehicle's accelerator would be termed a *first-order rate control*. The position of the control is related to the rate of engine output. The further the accelerator pedal is pressed the larger the request for more engine output. The shape of the engine's power curve, and other characteristics affecting the rate of power delivery such as engine specification, gear selection and vehicle weight, determine the time lag involved in enacting the driver's request. This is termed *system response*. In the case of vehicles the system response and associated time-delay can be characterised by exponential lag of varying degrees. In other words, the output does not exactly track the response of the input. Acceleration is not instantaneously related to accelerator pedal position, but vehicle feedback related to it often is. Depending on the vehicle, sensations that the engine is 'trying' to meet the driver's demand for greater acceleration can be felt near instantaneously, such as changes in engine note or feelings of acceleration.

As Figure 3.3 shows, in a low gear, at a low speed, pressing the accelerator will cause the vehicle to surge forward with little delay.

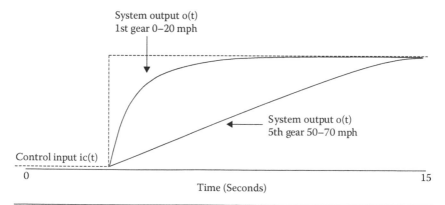

Figure 3.3 Transfer function for full acceleration in first gear to 20 mph and in fifth gear between 50 to 70 mph.

The time lag here, though still exponential, is relatively small. In first gear a typical car will be able to deliver a target speed of 20 mph in around two or three seconds (Evans and Herman, 1976). In a high gear at high speed, pressing the accelerator will cause the vehicle to pick up speed much more gradually. A 20-mph increment in speed beginning at 50 mph and performed in sixth gear can take 10 seconds or more.

The accelerator can also, in some circumstances, be understood as a second-order acceleration control. When seeking to rapidly attain a target speed, the driver may choose to press the accelerator pedal down much further than is needed to merely attain the target speed. Thus they are increasing the rate of change of vehicle acceleration; this is demonstrated in Figure 3.4. Once the target speed has been attained the driver then releases the accelerator back to a position commensurate with the target speed rather than acceleration towards it. The tracking function thus involves imparting an acceleration towards the target speed, and the rate of this acceleration has to cease once the target speed is reached.

3.2.3 Driver Feedback and Powertrain Design

Transfer functions enable the human input to be mathematically related to the machine output. They also enable us to explore further what changes in command input values $ic(t)$, sources of error $e(t)$ and the force applied to the relevant control $f(t)$ and its movement $f(t)$,

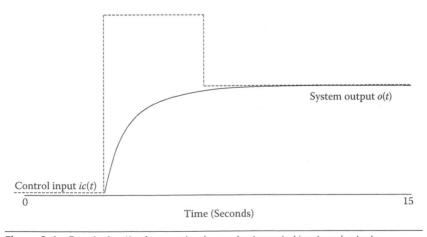

Control input $ic(t)$

0   System output $o(t)$ 15

Time (Seconds)

Figure 3.4 Transfer function for second-order accelerator control inputs and outputs.

along with overall control $u(t)$ and system dynamics $o(t)$, mean for driver cognition and behaviour. The focus of interest is how feedback $id(t)$ links these different elements together, particularly as Rumar (1993) points out humans have evolved to act directly upon their environment and thus receive direct, instantaneous feedback from it. We can return once more to 1978 to appreciate what direct feedback and interaction might look like from an automotive engineering perspective.

In 1978 pressing the accelerator pedal in a Mini 1000 would result in the following. The descending pedal would pull a thin steel Bowden cable through its outer casing, pulling open a throttle plate inside the A-Series engine's SU carburettor. The partial vacuum being created by the pistons inside the engine would now be able to draw in more air through the air filter. The partial vacuum would also cause a spring-loaded piston in the carburettor to lift up. This would enlarge a 'fuel jet' and allow a corresponding increase in the amount of petrol being added to the increased airflow. This fuel–air mix would enter the engine's cylinder head where it would be ignited by a spark plug, the hot expanding exhaust gases would exert more force on the pistons, and more power and speed would be generated contingent on the engine's inherent power curve. There is a very direct relationship between driver inputs and the fundamental volumetric and power characteristics of the engine.

In 2018 pressing the accelerator in a new Mini One results in the following. An electrical signal is sent from a potentiometer in the accelerator pedal to the engine control unit (ECU). The ECU will compare the value received from the accelerator pedal with the value currently being signalled by the drive-by-wire throttle. The ECU will also calculate the driver's demand based on how far and fast the pedal has been pushed, and reference a host of other operating parameters to decide on the optimum engine configuration to meet those demands. These include the current road speed, gear, valve timing and lift settings, turbocharger boost pressures, intake and engine temperatures, the current air mass being consumed by the engine, and the oxygen content of the exhaust gases (among numerous others). The ECU will then take these inputs and refer to a look-up table or 'engine map' and select the configuration needed to reach the new driver-requested state. Based on this the ECU will open the throttle – often by considerably more than the accelerator input would suggest – in order to bring the turbocharger up to optimum speed and boost quickly. All of these previous readings will be constantly referenced in order to continually adjust the engine's operating configuration, including further computer controlled changes in throttle position, until a new demand is made by the driver.

This is a very indirect relationship between driver inputs and the fundamental volumetric and power characteristics of the engine which the computerised control system now effectively masks. Indeed, while speed control in a car – from the driver's point of view – is characterised by a particular transfer function, behind the scenes there will be many other higher-order, and more difficult to control, functions which the technology is in charge of managing. The accelerator has perhaps become less of a 'control' and more of an 'interface'.

The situation with hybrid drivetrains is even more complex. The very different power curves of two quite distinct 'prime movers' are now having to be integrated and behave 'as one'. In this case computerised control is needed so that internal combustion and electric power is deployed in ways that are mutually reinforcing but also seamless. The issues with full electric drivetrains are equally interesting. The possibilities arise not only to provide extremely 'timely' feedback due to very high torque capabilities but, in theory at least, computerised control can bestow upon an electric drivetrain a very wide range of different power curves. There is no particular technical barrier for a new Mini E to precisely mimic the power curve of a 1978 Mini 1000, or indeed a 1978 Jaguar XJ. In fact, the new Mini E would probably be faster.

In all these cases the accelerator pedal behaves as a first and second order rate control regardless of the underlying technological sophistication. Indeed, the main reason for much of the sophistication is so that precisely these 'ergonomic' properties remain. If they were not present then pressing the accelerator pedal would lead to highly uncertain outcomes: in some conditions the car would be highly responsive, in others it would be sluggish. There would be pronounced peaks and troughs in the engine's response to pedal inputs and it would be extremely difficult for a driver to maintain a steady cruising speed. Early turbocharged cars and the troublesome phenomenon of 'turbo lag' provided a perfect demonstration of the problems. Pressing the accelerator in early turbocharged cars would often lead to little imme-diate increase in speed. This was because of low engine compression ratios, needed due to a lack of sophisticated fuel injection, and a lack of low-speed turbo boost, due to the comparatively crude turbo-charger technology. This would inevitably lead the driver to press the accelerator harder, upon which the engine's exhaust gases would cause the turbocharger to spool up; there would be an abrupt rise in engine torque and a manic screech of the tyres as they were overwhelmed by the power, followed by an explosive thrust towards the horizon. Hardly an ergonomic solution, and one that modern-era electronics disguises almost completely.

3.3 Transmission Systems

3.3.1 Gearbox Basics

Cars need gearboxes – at the moment at least – because of the way internal combustion engines deliver their power output. These engines need to be turning at a given speed in order to generate enough power to move the vehicle. In addition, the range of engine speeds over which engines deliver their optimum power and torque is fairly narrow compared to the accelerative needs of drivers. Gearboxes and transmissions enable the engine to offer the car brisk acceleration from standstill, rapid overtaking capabilities, and the ability to main-tain a high cruising speed at a relatively relaxed engine speed. One of the attractions of full EV drivetrains is that no such complication is needed. Electric motors have a much wider power/speed range com-pared to reciprocating piston petrol/diesel engines.

Conventional manual gearboxes are a purely mechanical design with an ancestry dating back 100 years when a gearbox was actually shaped like a box and filled with intermeshing gear wheels (Setright, 1999). Decades of development, including the 'synchronizer' intro-duced by General Motors in 1928, improved refinement considerably (Nunney, 1998). Today the output from the engine passes through a clutch mechanism which enables the drive from the engine to be disconnected from the gearbox. This is necessary for the synchro-nizer hubs inside the gearbox to be able to select the appropriate gear ratio to output to the drive wheels. A mechanical linkage from the gear lever runs directly to the gearbox and operates the synchronizer mechanisms. The gear lever and linkages operate the synchronizer hubs, and it is these devices which physically change the gears by locking the otherwise freewheeling cogs to the output shaft. This is shown in simplified form in Figure 3.5.

3.3.2 Gear Changing as a Tracking/Manual Control Task

From the driver's point of view the manual gear change process is threefold. First they have to disconnect the engine from the gearbox using the clutch. They then have to change the gear, and then reconnect the engine by raising the clutch pedal. There is no delay in the action upon the clutch pedal and the desired consequence of actually disengaging the clutch. The clutch mechanism, therefore, is a case approximating to zero-order direct tracking. It is approximate because the clutch mechanism is also progressive, not like an on/off switch. For some driving tasks, such as changing gears, the clutch pedal has to be fully depressed; for reversing or slow manoeuvres the pedal has to be held somewhere between the fully raised and fully depressed position. Fortunately, drivers seem to be readily able to finely adjust the position of the pedal to achieve the desired amount of power transmission to the road wheels, and are quite sensitive to the resistance of the clutch pedal and clutch mechanism in doing so (Southall, 1985). This

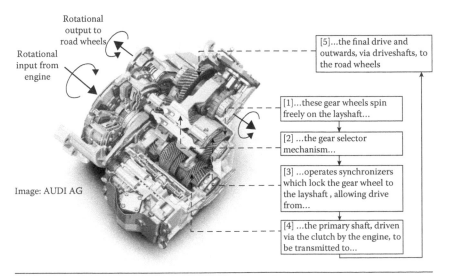

Figure 3.5   Gearbox cross section and principles of operation.

process is also intimately bound up in the feedback provided to the driver. As the engine is subjected to load by raising and progressively engaging the clutch, its speed and engine note will fall. This is a cue to apply more power by pressing the accelerator. Similarly, the smoothness of forward velocity (combined with auditory feedback) helps the driver to determine what position the clutch pedal should be in. The transfer function in Figure 3.6 depicts the relationship between control input in the form of clutch pedal position, and system output in the form of power output to the driving wheels.

3.3.3 Driver Feedback and Transmission Design

A favourable aspect of manual gearboxes is that the direct mechanical linkage provides abundant cues for accurate gear selection. Firstly, (gear) levers 'afford' pulling and pushing (Norman, 2013), although in exactly what direction takes time to learn. Gear changing is a complex psychomotor task, demonstrated by learner drivers who need to look down at the gear lever while operating it (Shinar, Meir and Ben-Shoham, 1998). Drivers rapidly acquire experience and the activity becomes skilled, quick and easy to perform. Even then, drivers can still be informed by the transmission of further possible errors in gear selection. This occurs through the audible clashing of synchronizer engagement mechanisms if the clutch is not depressed while attempting to change gear, or, if the desired gear is trying to be engaged at too high a speed then gear lever resistance becomes very high. These examples illustrate how the direct mechanical linkage provides a range

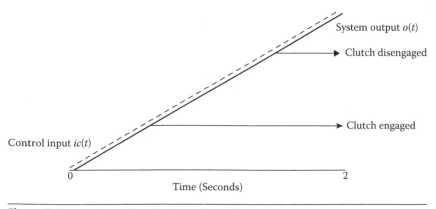

Figure 3.6 Transfer function relating clutch pedal inputs and system outputs.

of useful affordances and constraints which make gearbox operation relatively transparent. The driver is not likely to know the technical operating characteristics of layshaft constant mesh gearbox mechanics but they can construct a fairly accurate mental model of its behaviour. This helps ensure the designer's and user's mental models of system operation are compatible despite being very different (Norman, 1998).

Already it has been shown how vehicle acceleration can be specified in terms of first- and second-order rate and acceleration control. The gearbox plays an important further role in terms of overall vehicle system dynamics. Not only will the driver depress the accelerator more than is necessary to maintain the target speed (in order to reach that speed quicker) but they may also increase the rate at which a target speed is acquired by changing down a gear. The gearbox, therefore, serves to alter the overall gain of the system.

A relatively recent trend is an increase in the number of gears or 'system gain intervals' now provided (Table 3.1). It probably will not surprise most readers to learn that our 1978 Mini example from earlier has just four gears or four discrete levels of system gain. In fourth gear the gain equates to 16.5 mph for every 1000 rpm of engine speed, making a 70-mph motorway cruise a very noisy (by modern standards) 4200 rpm. This contributes greatly to the enhanced perception of speed. Such a high cruising engine speed also meant that further increases in speed "becomes very harsh and noisy" and if "you try to use high[er] revs and full throttle it becomes fussy and breathless and does not seem to go that much quicker" (Motor, 1979, p. 39). This is because the engine's peak power output occurs just a little higher than the 4200 rpm cruise at 4750 rpm. After this point the power curve begins to fall and the accelerator behaves less like a first-order rate control. Increases in pedal pressure do not yield significant further increases in speed or acceleration. A similar sensation can be recreated by driving a 2018 Mini One at 70 mph in third gear.

While it may not be much of surprise to learn that a 1978 Mini had just four gears, it is perhaps surprising to learn that our 1979 Porsche 911 Turbo also had just four gears. In the case of the Porsche the system gain intervals (i.e. gear ratios) were considerably different. 70 mph in the Porsche, for example, required just 2630 rpm in fourth gear. What is interesting is the new Mini One, which is far from atypical, has an even higher top-gear ratio. Its

Table 3.1   Comparison of System Gain for Different Vehicles

SYSTEM GAIN INTERVAL	MAXIMUM SPEED		
	1978 MINI	1979 PORSCHE 911 TURBO	2018 MINI ONE
1st Gear	30 mph	48 mph	35 mph
2nd Gear	47 mph	83 mph	60 mph
3rd Gear	73 mph	123 mph	91 mph
4th Gear	82 mph[a]	160 mph	109 mph
5th Gear	n/a	n/a	121 mph[a]
6th Gear	n/a	n/a	121 mph[a]
Engine Revs at 70 mph	4200 rpm	2630 rpm	2300 rpm

[a] Limited by available engine power.

engine speed at 70 mph in top gear is just 2300 rpm. A 4200 rpm top-gear cruise in a modern Mini (one powerful enough) would in fact be over 130 mph. If power was unlimited then at the engine's maximum speed of 6500 rpm the Mini One would be travelling at 201.5 mph. Clearly this is not the design goal. The goal, as shown in the bottom row of Table 3.1, is to reduce engine speeds at 70 mph (by nearly half compared to the 1978 Mini), reduce noise (from an excruciating <90 dBA to a much more modest 67 dBA) and increase fuel economy (from 33.1 mpg to 60.1 mpg). All this is achieved by placing cruising speeds at the point at which the Mini's small turbocharged engine is able to deliver maximum torque.

3.4 Braking Systems

3.4.1 Braking System Basics

The vehicle's engine, helped by its gearbox, is responsible for positive longitudinal acceleration. Aerodynamics, rolling resistance, and engine braking all exert the opposite retarding force. In stopping the vehicle the brakes often generate a force 10 times greater than that used to put the vehicle in motion. Cars can decelerate much faster than they can accelerate. Vehicle brakes, like engines, are essentially energy converters. This time they turn the kinetic energy of the moving vehicle into heat energy (Schwaller, 1993). The faster and more efficiently kinetic energy can be turned into heat, the better the braking system.

Modern passenger cars employ a hydraulic braking system which offers a very large mechanical advantage to the driver. Although drivers are capable of pressing the brake pedal with a force often in excess of 1000 Newtons, this same force would have little or no braking effect if it was the sole of the shoe being pressed straight on to the tarmac (Mortimer, 1974). To multiply this force the foot pedal operates a brake master cylinder that pressurises hydraulic fluid. This is also air assisted via a vacuum servo unit. The hydraulic fluid runs, via pipes, to cylinders or pistons that press a fixed stationary friction material onto rotating friction linings attached to the vehicle's wheel hubs. A normal braking system enables the driver to pressurize the system to as much as 2000 pounds per square inch, with a foot pedal force of perhaps only around 400 Newtons or less (Mortimer). This enables the friction materials to be pressed against the friction linings with many tonnes of force. Consequently, friction temperatures can rise extremely rapidly. The challenge is to get rid of this heat as quickly as possible.

3.4.2 Braking as a Tracking/Manual Control Task

Modern braking systems typically employ a metal disc as the rotating friction lining and metal/organic compound stationary friction materials mounted inside a form of hydraulic clamp or brake calliper. Disc braking systems were introduced by Chrysler in 1950 and are now widely used (Newcomb and Spurr, 1970). In conjunction with a powerful hydraulic system a very high clamping force on the rotating disc can be achieved, coupled with efficient and stable heat dissipation.

This braking force is transmitted to the road via the vehicle's tyres. As long as they possess sufficient grip on the road surface then braking performance is approximately linear, with the brake pedal assuming the characteristics of a zero-order position control. This means movement of the control inceptor controls the output of the system in a direct relationship, as shown in Figure 3.7. Although the initial phase of braking does involve a transition state, in which the vehicle's weight distribution changes and its front suspension compresses, within the normal envelope of operation there is no time delay in pressing the brake pedal and the resultant deceleration beginning.

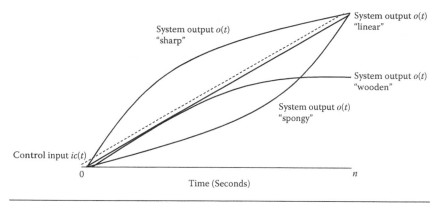

Figure 3.7 Transfer function for zero-order braking system, showing stylized functions for sharp, spongy and wooden feeling brakes.

3.4.3 Driver Feedback and Braking System Design

Despite this apparently straightforward input/output relationship, a host of adjectives are used to describe the feel of vehicle brakes. Our 1978 Mini brakes are described as "err[ing] on the heavy side [...] do feel rather dead" (Motor, 1979, p. 39) while the Porsche 911 Turbo's were also "...a little heavy, but they were beautifully progressive and very reassuring, imparting plenty of information to the driver" (Motor, 1979, p. 226). Harris and colleagues (2005) have identified approximately 400 such adjectives used within the motoring press, all of which arise because of subtle deviations from a strictly zero-order control system which drivers are sensitive enough to perceive as vehicle feedback. For brakes, some notable adjectives are translated into transfer functions in Figure 3.7.

Outside of normal operational parameters the direct zero-order relationship begins to falter. If the clamping force of the friction materials onto the friction lining overwhelms the amount of grip available between the tyre and road, then the road wheel will lock. Maximum braking performance actually occurs, to all intents and purposes, at the instant just prior to the wheels locking and the tyres skidding. This is what anti-lock braking systems (ABS) achieve by pulsing the brakes off and on at full power very rapidly (Schwaller, 1993). Figure 3.8 illustrates the consequences of brake pedal input and output when the generally first-order relationship starts to collapse.

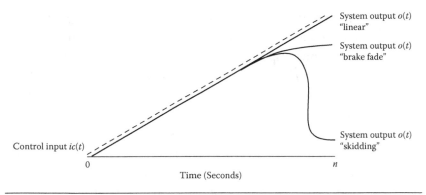

Figure 3.8   Braking input versus output showing normal braking, skidding and brake fade.

The other instance of the direct zero-order relationship faltering is due to a phenomenon termed *brake fade*. If the heat generated by the brakes reaches a certain level, and cannot be dissipated quickly enough, then brake pedal effort increases whilst braking performance decreases (the brakes feel 'wooden'). Brake fade occurs due to the braking components not being able to dissipate the heat generated by the vehicles kinetic energy, and this in turn can cause the hydraulic fluid which supplies the braking force to the callipers to boil and/or the friction materials to catch fire. With the brakes in this condition the direct linear relationship between braking force and brake pedal position fails. Brake fade becomes counterintuitive to the driver who invariably will press the brake pedal even harder. In a situation not dissimilar to sudden unintended acceleration the feedback loop reverses. In this condition the brake needs to cool down, and continuing with a brake application will achieve the reverse of what the driver is demanding. Thus the situation becomes that of unstable 'positive feedback', or continuing with the very activity that only serves to make the error worse.

3.5 Steering Systems

3.5.1 Steering Basics

The systems responsible for controlling the longitudinal performance of the vehicle are operated by pedals pressed by the driver's feet. The steering system controls lateral direction, and is operated by the hands and arms using a steering wheel. The palms of the hands are one of the most touch-sensitive parts of the human body.

The driver imparts movements to the steering wheel ($f(t)$) which initiate lateral accelerations of the vehicle. Although highly dependent on steering and suspension geometry most steering systems will see operating effort change according to speed (Bashford, 1978). At lower speeds the steering effort is generally higher although lateral G-forces resulting from steering manoeuvres are lower. Inversely, at high speeds, steering effort generally reduces, but steering actions give rise to much higher lateral accelerations. Thus it can be said that increasing vehicle speed serves to increase steering system gain. As such, the level of lateral accelerations acceptable to drivers tends to decrease with speed (Lechner and Perrin, 1993; Ritchie et al., 1968; Winsum and Godthelp, 1996).

3.5.2 Vehicle Steering as a Tracking/Manual Control Task

The steering wheel controls the acceleration of the vehicle towards its desired lateral trajectory (Sanders and McCormick, 1993). Vehicle steering is an example of second-order acceleration control. When changing lane on a motorway, for example, the driver has to impart a small lateral acceleration then return the wheel to the centre position or null point. As the vehicle traverses the adjacent lane markings the driver then has to impart another lateral acceleration in the other direction to enable the vehicle to assume its position within the travel lane. When this is achieved the driver returns the steering to the centre. In terms of transfer functions vehicle steering is also a case of exponential lag. The time lag is only very slight and arises from the characteristics of pneumatic tyres and front suspension geometry (Figure 3.9).

Humans are noted for being relatively poor at tracking tasks such as that described by vehicle steering (Sanders and McCormick, 1993). One of the main problems in second-order acceleration tracking relates to returning to the null/centre point after initiating an acceleration of some form. The analogy is one of tracking a target on a computer screen using a joystick, in which the user may repeatedly overshoot the desired point before converging on it. Fortunately, vehicle steering has a natural tendency to return to the centre null point, and this self-centring action greatly facilitates tracking performance. This feature arises primarily from the geometry and design of the vehicle's front

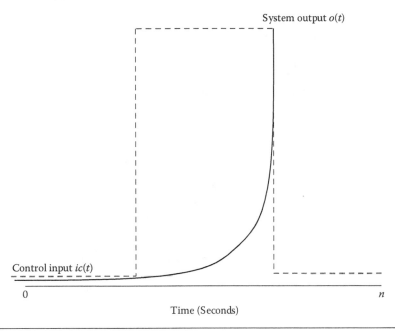

System output $o(t)$

Control input $ic(t)$

0

n

Time (Seconds)

Figure 3.9 Transfer function for vehicle steering.

suspension system. Parameters which affect self-centring are king-pin inclination, castor angle and camber angle (Hall, 1981). A detailed technical explanation of this is not required here, but Figure 3.10 illustrates the key dimensions of interest.

3.5.3 Driver Feedback and Steering System Design

Since the 1970s most European cars use a rack and pinion steering mechanism, illustrated in Figure 3.11. The reason steering has to be based on gearing and rotating steering wheels is because cars are heavy. There has to be some mechanical advantage in order for the driver to be able to physically turn the steering wheel. This advantage is often quite large. It is not unusual for the steering wheel to travel about four metres from lock to lock in order to turn the road wheels in an arc of only about 90 degrees (Setright, 1999). As cars have become heavier the need has arisen to increase this mechanical advantage by deploying either hydraulic or electric assistance.

Rack and pinion steering is a particularly simple and direct method of steering the front wheels and is widely used. It also permits access for feedback from the road to travel back up the system to the steering

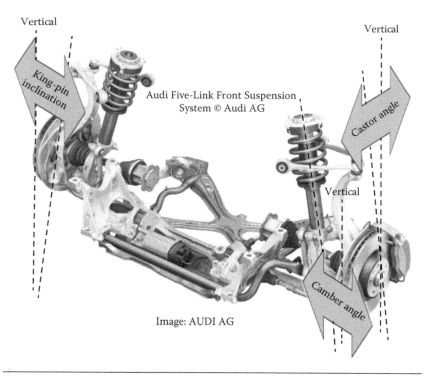

Figure 3.10 Front suspension geometry: key parameters include king-pin inclination, castor angle and camber angle.

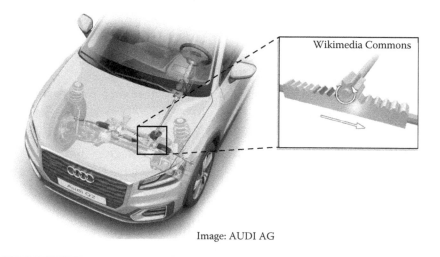

Figure 3.11 Rack and pinion steering system.

wheel, where it is perceived by the driver as force feedback. This is not the case with all technical solutions, some of which are 'irreversible' and uni-directional. 'Worm and roller' steering (Figure 3.11), for example, only permits forces to be transmitted from the steering wheel to the road, not in reverse. Where it is permitted, steering feel arises partly from disturbances involving the road surface (e.g. bumps, cambers, cornering forces etc.) and stored energy (or springiness) in the vehicle's tyres and other compliant parts of the steering system (such as rubber mountings etc.). Steering feel also arises out of 'aligning torque', or the effort required by the driver to hold the steering wheel in its desired position. The more aligning torque, the more cornering force is developed by the vehicle's tyres (Jacobson, 1974). In other words, the harder and more extreme the cornering manoeuvre, the more aligning torque is needed to hold the steering wheel in that position. This is an important determinant of vehicle steering control dynamics.

Joy and Hartley (1953/4) describe aligning torque as giving "a measure of the force required to steer the car, i.e. it gives a measure of the 'feel' at the steering wheel" (p. 113). This feel arises due to a characteristic of pneumatic tyres termed *pneumatic trail* (Setright, 1999). This is a speed- and load-sensitive distortion of the tyre's contact patch with the road which is fed back through the steering system. Pneumatic trail, and how this interacts with the geometry of steering and suspension components, is responsible for the weight and force needed to operate the steering wheel (Joy and Hartley, 1953/4). Up to a point, cornering force versus aligning torque is almost directly proportional, meaning it is possible to corner faster by imparting more effort to hold the steering wheel in a desired position. When this linear relationship between cornering force and aligning torque starts to falter, i.e. it is possible to corner harder and faster with no further increase in steering torque, most drivers perceive this to be the point at which control is about to be lost. Strictly speaking, as Figure 3.12 shows, it isn't.

There is a 'dead zone' where cornering forces can still build but steering effort no longer increases, and this can be seen as the flattening of the aligning torque curve. One of the skills of racing drivers is to bring the vehicle into this 'dead zone'. When steering effort starts to fall away rapidly the transition from the dead zone to loss of grip has begun. The challenge, therefore, is to ensure a gradual rate of state change, from a proportional relationship into a gentle transition to the dead zone and, again, gradually into loss of grip. Self-aligning torque

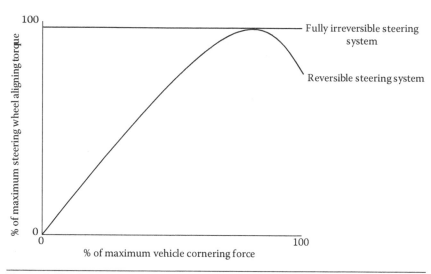

Figure 3.12 Relationship between cornering force and aligning torque supplied by the driver.

and providing sensitivity to changes in the road surface have been cited as important design goals. Drivers need this rapid indication of environmental state change in order to control the vehicle effectively (Joy and Hartley, 1953/4). The shape of the aligning torque versus cornering force curve is very important in defining the control dynamics of the steering system. Indeed, very few manufacturers deviate from a steering system which is self-centring and has a rising speed/aligning torque relationship despite the fact the vehicle itself needs no such properties. To do so would, in the words of one automotive engineer, induce 'bafflement' for the driver (Jacobson, 1974). This has not stopped several worthy attempts being made (Figure 3.13).

An interesting example dates from the 1970s in the form of the DIRAVI (*Direction à rappel asservi*) system fitted to various Citroën models. This was a fully hydraulic power steering system that was not only very highly geared (approximately one turn of the steering wheel to go from lock to lock) as well as strongly and artificially self-centring (even with the car stationary), but also 'irreversible': there was no feedback from the road wheels to the steering wheel. Engineers were well aware that having such a highly geared steering system would be difficult for drivers to control at high speed, so the system was speed sensitive, becoming much 'stiffer' to operate in proportion to road speed. This did not stop reviewers noting that: "We still find the [...] steering ergonomically unsound, though, for its extreme directness demands

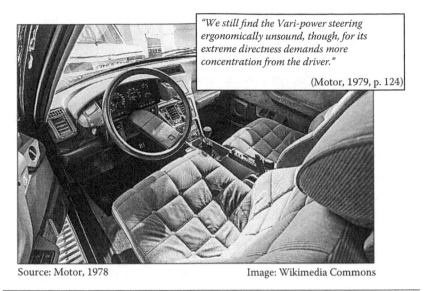

"We still find the Vari-power steering ergonomically unsound, though, for its extreme directness demands more concentration from the driver."

(Motor, 1979, p. 124)

Source: Motor, 1978   Image: Wikimedia Commons

Figure 3.13   The irreversible hydraulically powered Citroën DIRAVI steering system.

more concentration from the driver than conventional power steering systems" (Motor, 1979, pp. 124–125).

Readers may also recall the Eureka PROMETHEUS project which ran from 1987 to 1995. One of the outputs of this was a Saab 9000 with an Airbus-type joystick controlling the steering. If nothing else it revealed the physical ergonomic benefits of steering wheels in reducing driver (wrist) fatigue and improving control accuracy, not to mention drivers reporting a disconcerting lack of feedback. While the joystick interface was, in the words of Saab, "a project which remains at the level of research" the underlying drive-by-wire technology is currently alive and well in the present-day Infiniti Q50.

This vehicle has full electronic steer-by-wire, albeit one that still uses a steering wheel as a control inceptor. The benefits of such a system are numerous: a fully electronic system with no mechanical backup is cheaper and lighter, it makes the interior more crashworthy, it interfaces much better with advanced driver assist systems, and it creates new and interesting possibilities for different types of control inceptor and feedback. Far from noting the attractive feature of this technology, reviews highlight it as a particular problem: "the steering, which offers next-to-no feedback, mak[es] it difficult to put your trust in it. [...] You can also change the amount of resistance and the quickness of the steering, but making it any faster than the standard mode can make the car feel really twitchy" (*Auto Express*, 2017).

Most current power steering systems are not like any of these previous systems. They utilise a manual rack and pinion steering system which connects the steering wheel mechanically and directly to the front wheels. The direct mechanical system is 'assisted' either by a hydraulically powered ram or an electric motor. Despite preserving a direct mechanical link, assisting the steering, of course, influences the shape and contour of the self-aligning torque curve and places a further 'intervening variable' between the driver and their environment. We have already seen how important this is in Chapter 2 and it also emerges as the leading 'subjective rating' in Harris' research on the adjectives used to describe vehicle handling variables (2005). Steering can be communicative/vague; informative/uninformative; precise/imprecise; sensitive/insensitive; and well-weighted/poorly-weighted. A purely technically-optimised solution certainly does not guarantee favourable driver perceptions let alone situation awareness (SA).

3.6 Chassis Design

3.6.1 Chassis Basics

The driver controls the longitudinal and lateral directions of the vehicle using the engine, the brakes and the steering. The demands made by these active systems have to be dealt with by passive vehicle systems such as the vehicle's suspension and its chassis or superstructure. A number of interrelated factors affect the way in which the vehicle responds to the demands made on it by the driver. The distribution of vehicle mass, centre of gravity, design of steering, braking, powertrain and suspension systems define how the vehicle will respond to driver inputs and how it will behave under dynamic conditions (Jacobson, 1974).

3.6.1.1 Rear-Wheel Drive The basis of a modern car is its floorplan. The layout of the vehicle's primary mechanical systems around this floorplan plays a large part in defining its dynamic behaviour. Historically the drivetrain of road vehicles has been longitudinal, with the engine mounted at the front of the car and the gearbox mounted behind, with the gear stick appearing straight out of the gearbox by the front seats. A propeller shaft connected the drive to the rear axle that, via a differential gear, turned the drive through 90 degrees to the rear wheels.

There are various advantages and disadvantages associated with this layout. For larger cars, and certainly for rear-engine cars, traction tends to be good, and there are often favourable implications for weight distribution as these heavy driveline components, such as engines and gearboxes, are often distributed more evenly across the length of the vehicle. As a result, front and rear wheels bear a more or less equal share of the vehicle's total weight. With more even weight distribution comes the possibility to adopt a less compromised front suspension geometry. Instead of being optimised to counter the effects of weight distribution and drive to the front wheels, it can be optimised for self-centring and favourable camber changes in cornering. Although this might cause steering effort to increase, it still remains within acceptable boundaries due to more favourable weight distribution to start with.

3.6.1.2 Front-Wheel Drive Front-wheel drive is a more recent development which has become the de facto standard for the majority of today's passenger cars. Approximately 85% of cars on British roads are front-wheel drive. The layout was made possible by the invention of the constant velocity joint by Czechoslovakian engineer A. H. Rzeppa (Nunney, 1998). This is a form of flexible joint which enables power to be transferred to the front wheels even though they move up and down on their suspension, and also swivel from side to side when steering. One of the first passenger car applications of front-wheel drive was the Mini of which our 1978 version shares a direct lineage. The later Fiat 128 refined the concept into something approaching the current configuration. Unlike the Mini it led modern practice in having the gearbox mounted on the end of the engine, rather than underneath it.

Front-wheel drive can provide good traction (especially for smaller vehicles) and compact vehicle packaging solutions by removing the need for a bulky propeller shaft, differential and driven rear axle. The disadvantages can be unfavourable forward weight distribution as large as 70/30 in some cases, and acceptance of certain other engineering compromises now that the front wheels of the vehicle have to deal with power transmission, steering and a large proportion of vehicle braking.

These compromises can invoke problems with steering force feedback, self-centring and aligning torque. Firstly, as the vehicle traverses bumps and road irregularities, the angular disposition of the flexible driveshaft running from the transmission to the front wheels is unbalanced between the left and right front wheels. This makes the drive torque unequal, meaning front-wheel drive cars can tend to turn gradually to the outside of a corner when the accelerator is depressed, and inside when the accelerator is released (Gillespie and Segel, 1983). Secondly, with so much of the vehicle's weight now over the front wheels, steps need to be taken to ensure steering effort is not excessive. King-pin inclination needs to be minimised as this can greatly increase steering effort (the front wheels would have a tendency to lift the car otherwise). Doing this, unfortunately, can rob the steering of its self-centring behaviour (Bashford, 1978). Thirdly, front-wheel drive vehicles also tend to have very small castor angles (a little like having vertical dead-straight forks on a bicycle). This further reduces steering effort, but on its own can induce stability problems and excessive sensitivity to the road surface, which can combine negatively with the unequal drive torque to promote a phenomenon called *torque steer*. This is when the vehicle can pull or tug in a lateral direction when the accelerator is pressed. To overcome this, amongst other solutions, techniques such as having proportionately more of the wheel width to the inside of the wheel centre line (called *negative offset*) and parallel steer characteristics (equal turning angle of the front wheels even though they travel through unequal arcs) are applied to restore stability. This latter feature can be why a car will appear to 'scrub' with its tyres squeaking when tightly manoeuvring in a car park, for example. But again, these aspects are disadvantageous to feedback and self-centring (Pitts and Wildig, 1978). This all serves merely as an example of the subtle decisions and compromises the automotive engineer has to make, and they can vary greatly between different vehicles (Figure 3.14). The effect, however, is to influence once again the shape of the transfer function and the feedback/control dynamics of vehicle steering, and not always in a desirable way from a Human Factors perspective.

3.6.1.3 Four-Wheel Drive   An increasing number of passenger vehicles now come equipped with four-wheel drive. In the past this would often rely on mechanical 'Torsen' style differentials to apportion drive torque between the two driven axles, a solution which provided

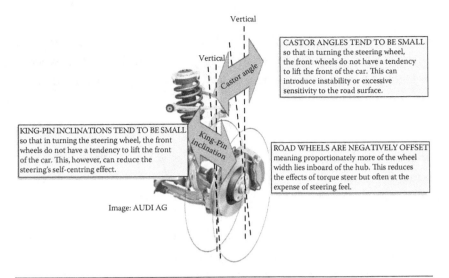

Figure 3.14 Front-wheel drive design parameters and their impact on driver feedback.

excellent traction at the expense of fuel economy. Such systems also tended to require a longitudinal, rear-drive layout which, as noted above, is increasingly uncommon. A more recent development is the Haldex-type system. This is used to provide four-wheel drive capability across a wide number of normally front-wheel drive vehicles. It is the 'Quattro' system as offered on many Audis, for example. It is no coincidence the trend towards higher power densities and outright power outputs has coincided with the trend for four-wheel drive. At the time front-wheel drive was introduced in the Austin Mini and other similar vehicles, there was insufficient power for problems of traction and torque steer to become seriously manifest. With power outputs for some front-wheel drive cars exceeding well over 200 bhp these issues can be acute.

Haldex-type four-wheel drive systems rely on increasing amounts of electronic control, including an electronically controlled clutch mechanism to divert drive torque to the front or rear of the vehicle. Many current four-wheel drive cars have their drive biased (sometimes quite strongly) to the front, with drive torque only meaningfully applied to the rear wheels under more extreme conditions. This is combined with selective wheel braking to provide a function similar to the more mechanically complex limited-slip differential. All-wheel drive, combined with advanced handling management, offers vehicle designers great scope for resolving some of the compromises inherent in two-wheel drive platforms. A shift to electric power makes the prospect of all-wheel drive even easier to contemplate, with small individual motors providing power to individual wheels, or electric power driving one axle and fossil-fuel power the other.

3.6.2 Vehicle Control as a Tracking/Manual Control Task

The chosen layout and configuration of a vehicle's drive line components describes a significant portion of its behaviour when it in motion. A car traveling in a straight line will continue to do so unless deflected or slowed by external forces, represented within the tracking loop as $id(t)$, the disturbance input, and $ic(t)$, the control input (Wickens, 1992). Once in motion the car is said to have inertia. Steering the vehicle around a turn creates a *moment of inertia,* a term which refers to masses rotating around an axis (Meriam and Kraige, 2014). Theoretically speaking, when the driver initiates a steering manoeuvre they do so by traveling in an arc around an imaginary axis at the centre of the arc (Figure 3.15). The tyres, being

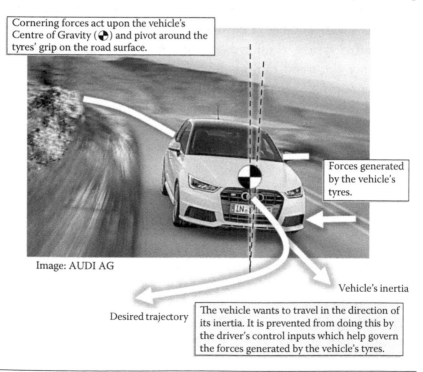

Cornering forces act upon the vehicle's Centre of Gravity (◕) and pivot around the tyres' grip on the road surface.

Forces generated by the vehicle's tyres.

Image: AUDI AG

Vehicle's inertia

Desired trajectory

The vehicle wants to travel in the direction of its inertia. It is prevented from doing this by the driver's control inputs which help govern the forces generated by the vehicle's tyres.

Figure 3.15 Vehicle in steady-state cornering showing action of forces and inertia.

the main force-generating components, sustain the vehicle in this arc due to their friction with the road surface (Ellis, 1994). Turning and cornering manoeuvres create moments of inertia by setting the car in arcs described by the vehicle's tyres.

3.6.3 Driver Feedback, Chassis Design and Vehicle Handling

The driver, via the vehicle's steering, suspension systems and tyres, is working against the natural direction the vehicle's inertia wants to travel in. This conflict of forces is fed back to the driver as the car leans in corners and aligning torque increases. The vehicle's steering, suspension and braking systems are constantly in opposition with the vehicle's moving mass. The driver controls these mechanical systems and is therefore constantly working against these forces, and a little of the stress these mechanical systems are sustaining is fed back to the driver through the vehicle's behaviour (system dynamics $o(t)$) and controls (control dynamics $u(t)$).

In lateral manoeuvres, as long as there is adequate friction between the tyres and road the vehicle will follow the arc described by the road wheels. Pneumatic tyres always run with a small percentage of slippage (Ellis, 1994) so the arc is approximate and depends on a wide range of factors. Weight distribution, in combination with the disposition of the driven wheels (front- or rear-wheel drive), alters the relative angles of slip. If this slippage exceeds a certain point then friction is totally overcome. This can be initiated by steering too hard whilst traveling too fast. Under this condition the vehicle will lose its moment of inertia and revert to some other instantaneous direction of velocity. This will be approximately tangential to the imaginary arc described by the vehicle's old trajectory but modified by the relative slip angles present at each wheel due to weight distribution/power transmission etc. To put it crudely, a vehicle's handling characteristics can be specified by which set of wheels have the largest slip angles (Joy and Hartley, 1953/4).

If maximum cornering gives rise to the front wheels slipping more than the rear wheels, this is called *understeer* (Figure 3.16). In this condition the vehicle will tend to plough straight-on as the vehicle's inertia overcomes the front wheel's grip on the road surface before the rear wheels. Conversely, if the rear wheels lose grip before the

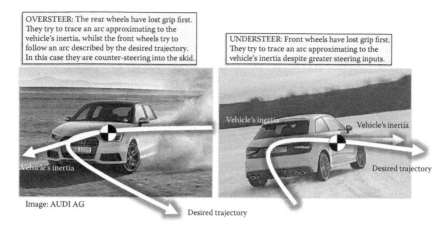

OVERSTEER: The rear wheels have lost grip first. They try to trace an arc approximating to the vehicle's inertia, whilst the front wheels try to follow an arc described by the desired trajectory. In this case they are counter-steering into the skid.

UNDERSTEER: Front wheels have lost grip first. They try to trace an arc approximating to the vehicle's inertia despite greater steering inputs.

Image: AUDI AG

Figure 3.16   Left-hand illustration shows an oversteering car, the right-hand illustration an understeering car.

front this is called *oversteer* (Figure 3.16). In this condition the rear of the car swings out wide. Understeer is a prominent characteristic of front-wheel drive vehicles due to the distribution of vehicle mass towards the front of the vehicle, and the fact the pool of total available grip at the front tyres is having to be shared among cornering, steering, traction, and around 70% of the vehicle's braking. This is not recommended, but an easy way to convert a heavily understeering vehicle into a potentially dangerous oversteering one is to drive it fast backwards, as *Autocar* magazine did with a Daewoo Matiz, turning it over in the process.

In a case of understeer the driver recovers the situation by slowing down and steering harder into the bend. With oversteer the driver has to correct the slide by steering in the opposite direction of desired travel, i.e. steering into the skid. These behaviours can also be captured by transfer functions. In a case of understeer, the vehicle retains the broad characteristics of second-order control except with much increased time lag. In the case of oversteer, the vehicle assumes the characteristics of a higher-order control level (third or fourth order perhaps). This is potentially challenging for drivers.

In a situation of oversteer the driver has to initiate corrective system accelerations by guessing at the correct amount required. Not just that, but the system is now sluggish and unstable. When the vehicle does respond it does so rapidly, and the driver has to make further guesses as to corrective actions. This is not easy for drivers and they typically do not cope well with a situation of oversteer (Godthelp and Kappler, 1988). As Figure 3.16 shows, under such situations control inputs, system outputs, and feedback effectively become uncoupled. The driver has lost control. For this reason understeering cars are generally regarded as safe-handling cars due to the retention of something approximating to second-order control. The downside is that front-wheel drive cars tend to run with larger slip angles at the front of the car and this subtly affects system response times by introducing lag. Time lag contributes to detriments in the quality of feedback (e.g. Welford, 1968) which brings us back, once again, to the Eurofighter Effect. Endowing a vehicle with larger slip angles at the rear reduces response lag and makes a car feel agile, while the electronic systems (when required) will step in to maintain a more benign form of second-order control if it looks like being lost.

3.7 Summary

A car embodies a chassis, suspension, steering, brake mechanisms, engine and transmission. All of these systems are under the control of the driver. 'Control' implies there is something to be controlled, and in the case of automotive systems it is a whole host of forces and opposing forces.

The driver imparts actions to the fundamental vehicle systems. These, in turn, transform driver inputs into outputs that work against opposing forces acting upon the vehicle's desired trajectory and speed. The fundamental vehicle systems place the driver at a large mechanical advantage to the environment but they nonetheless provide the driver with some sensation related to the forces at work on the particular system. This is vehicle feedback.

Manual control, tracking, transfer functions and system and control dynamics answer the question of what we are dealing with: in other words, what exactly a car is from a user-centred design and control point of view. Framed in this way it becomes clearer why vehicles, despite their complexity, are so well matched to drivers. Although humans can struggle with certain 'high-order' tracking tasks, cars redeem themselves by having a particularly direct first- and second-order tracking characteristic, along with continuous feedback and interaction. In most cases response times are rapid; where second-order control tasks are present they are aided by vehicle feed-back, and within the normal scope of vehicle performance a broadly linear relationship between input and output is sustained. What is also clear is how changes in automotive engineering design can subtly (and not so subtly) alter these control and system dynamics. If drivers are as sensitive as they appear towards vehicle feedback, then automotive engineering represents a powerful independent variable.

4

SITUATION AWARENESS REQUIREMENT ANALYSIS

4.1 Introduction

The cockpit of a 1978 Mini 1000 was a rather Spartan affair. One would sit on the upright deckchair patterned driver's seat, grip the bus-like steering wheel, and have an unobstructed view ahead due to the abundance of glass, a short bonnet, and slender crash-unworthy A-pillars. The compliment of driver displays in a Mini were also rather meagre by modern standards. A large centrally mounted Smiths speedometer dominated, flanked by hints of future failure modes in the form of complimentary temperature and oil pressure gauges.

How in-vehicle display technology has changed. It is becoming difficult to think of a modern car that does not feature a display screen of some sort measuring at least 170 mm across, and in modern Tesla vehicles the display is an enormous 430-mm touchscreen. Interestingly, they are often now positioned in roughly the same location as the original Mini's speedometer. How did we get from a minimalist compliment of analogue driver displays to a multi-functional 430-mm touchscreen? Who decided what information drivers now needed to see that warranted such a change? The answer is unclear.

As noted earlier, a concern among the user-centred professionals we work with at vehicle manufacturers is that new display technologies like these will continue to proliferate in an unstructured and random way, gradually filling up the available dashboard space and overwhelming the driver, who is expected to make sense of it all. Debates like this boil down to what the driver's actual needs are. In this context, then, situation awareness (SA) emerges as an extremely useful way to help define them. In this chapter a systematic and exhaustive SA requirements analysis is presented. The actual analysis was performed by Ipshita Chowdhury, a PhD candidate working in the field of SA and self-explaining roads (see Chowdhury, 2014; Walker et al., 2013). In exploring SA requirements we can also dig deeper into the underlying theory of SA because this greatly impacts on what is regarded as an SA requirement and how it is measured.

4.2 Methods for Extracting SA Requirements

4.2.1 The Individual SA Lens

Contained in the individual view of SA are three tacit assumptions:

- It is a cognitive phenomenon residing in the heads of human operators.
- There is an objective 'ground truth' of a situation available to be known.
- Good SA can be derived from reference to expert and/or normative performance standards.

These tacit assumptions lead to an SA requirement being knowledge about x, y and z which resides inside (and must be accessed from) the mind of the individual. With these tacit assumptions brought into consciousness the established method for deriving SA requirements, put forward by Endsley (1995), can now be described.

The first step is to decompose the task under analysis into its constituent goal-directed elements via a so-called Goal-Directed Task Analysis (GDTA). There is some controversy here because GDTA is similar to the much longer established Hierarchical Task Analysis (HTA), in widespread use since the 1970s and apparently not referenced in the former's development. The GDTA method, like HTA, begins with higher-level goals or outcomes to be achieved, and progressively decomposes them into their component operations. HTA has a number of major advantages over the proprietary GDTA method. Firstly, it is capable of representing (and fully specifying) goal timing and sequence. Secondly, it views goals as end-states rather than resources required by operators, the former arguably easier to define in a robust way. The third, and most significant, benefit of HTA is that it has a much better and longer-established pedigree. Given this, HTA remains our preference and a fully worked and validated HTA of driving is made available in Appendix 1. It can be put to immediate use, reuse and modification in any future driver SA requirements analysis.

Once decomposed using GDTA/HTA the SA elements relevant to each driving operation can then be defined. GDTA itself is anchored firmly to Endsley's three-level model of SA (Figure 4.1). As such, the second step of the SA requirements process – under the rubric of the individual SA view at least – is to categorise these elements into the appropriate SA Level (Level 1 – Perception; Level 2 – Comprehension; Level 3 – Projection). Thus defined, it becomes possible to design systems to meet these SA requirements, and test them to see whether the identified SA elements are actually present or not in the minds of users. This is often undertaken using probe-recall techniques such as SAGAT (Endsley, 1988) or SACRI (Hogg et al., 1995), an example of which follows in the next chapter. To summarise, then, viewed through the individual SA lens requirements are individual items of knowledge or information defined with reference to a task analysis, categorised into SA levels, and measured by detecting whether people can recall them.

4.2.2 The Team SA Lens

Early team SA applications involved scaling up individual models to the team level and incorporating a number of more specific team SA concepts. These included the degree to which every individual

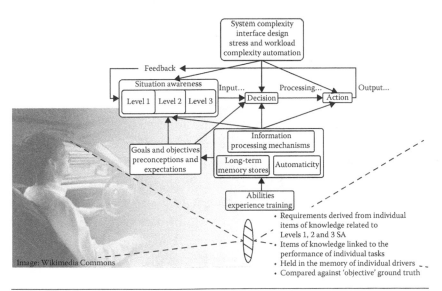

Figure 4.1   Endsley's three-level model underlies the individual SA lens.

driver, and those around her, possessed the SA required for their own individual activities within the traffic stream (e.g. Endsley and Robertson, 2000). It also included a fundamental requirement for shared SA, or 'the same SA' on mutual 'operational requirements' (Endsley and Jones, 2001). Using the three-level model as its basis, this approach to team SA would argue that participants in a wider traffic stream each have distinct portions of SA. Distinct in the sense it is related to their specific SA of their specific situation. In addition there is also overlapping or 'shared' portions of SA, or of things common to all participating drivers. It follows from this that successful 'team performance' needs individual drivers to have sufficient SA on their specific elements and the equivalent SA of shared elements (Endsley and Robertson, 2000). While this approach provides satisfactory answers to some types of team SA questions it is less satisfactory for others (Stanton et al., 2001). As a result, two team SA models have received particular attention: Salas et al.'s (1995) team SA model and Shu and Furuta's (2005) mutual awareness model.

The critical component of team SA, according to Salas et al., is information exchange. They argue that the perception of SA elements is influenced by the information exchange needed to accomplish task objectives and determine individual tasks and roles and the team's capability (i.e. expertise) alongside other team performance shaping factors (Salas et al., 1995). It is not simply shared or mutual awareness of operational requirements, but 'meta-awareness' of team processes too. On these terms information exchange can be viewed, in a simplistic sense, as communication between drivers in a road situation. In driving it would include information provided by vehicle indicators, the position and movement of other vehicles, non-verbal cues from other road users, the status of the wider driving environment, information from road signs and, of course, feedback from the vehicle. Salas et al. (1995) go further to argue that limitations in the knowledge brought to a situation by drivers, in the form of schemas, can be offset by information exchanged within the wider 'team' of actors and agents. For example, an inexperienced driver with no prior knowledge of a challenging road environment can simply follow a more experienced driver. In doing so they would create a better (and different) situation to be aware of.

They are benefiting from the cognition of the lead driver which is physically manifest in the driving behaviours they are emitting and making available to copy. Obviously, the same also holds true in the opposite sense with phenomena such as 'speed contagion' stemming from various perceptual and cognitive biases (e.g. Redelmeier and Tibshirani, 1999) along with team performance shaping factors such as conformity (e.g. Walton and Bathurst, 1998; Connolly and Aberg, 1993).

Under this view it is clear that individual SA underwrites team SA, which subsequently modifies that person's own SA in turn. What we have, in fact, is a cyclical process of developing individual SA and sharing SA with other team members, which then serves to modify both team and individual SA. This links to a different underlying model of SA: Smith and Hancock's (1995) perceptual cycle (Figure 4.2). In this model other drivers represent salient parts of the ambient surroundings. This modifies individual SA as well as providing an environment into which actions can be imparted and new situations created. Salas et al. (1995) states that team SA "occurs as a consequence of an interaction of an individual's pre-existing relevant knowledge and expectations; the information available from the environment; and cognitive processing skills that include attention allocation, perception, data extraction, comprehension and projection" (Salas et al., 1995, p. 125). Through

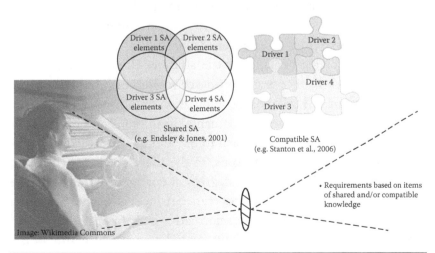

Figure 4.2   Smith and Hancock's Perceptual Cycle Model of cognition underlies team approaches to SA.

this lens SA becomes much more of an emergent property of inter-acting individual driver SA. What emerges is not of the same order of phenomenon as that of which it is composed. In other words, the collective SA residing within a traffic stream is not a simple summation of the individual drivers which comprises it (Fiore and Salas, 2004; Fiore et al., 2012). Shared SA, therefore, is not the same as 'identical' SA.

Shu and Furuta (2005) address this by proposing the concept of mutual awareness. This idea is conceptually similar to Endsley's shared SA model but with some important differences. Shu and Furuta argue that team SA comprises both individual SA and mutual awareness, that is, the mutual understanding of each other's activi-ties, beliefs and intentions. They further describe how team SA is a partly shared and partly distributed understanding of a situation among team members. In driving, for example, mutual awareness would be achieved when two drivers encountering each other are able to understand each other's behaviours and motives. An oncom-ing driver might flash their lights to indicate they are stopping and that the other driver can approach and pass. Or they might indicate and position their car to show they are making space. Or they might assert their road position in some other way to induce another driver to give way. And so on. Shu and Furuta (2005) would define this, and the resultant team SA, as "two or more individuals share the common environment, up-to-the-moment understanding of situa-tion of the environment, and another person's interaction with the cooperative task" (Shu and Furuta, 2005, p. 274).

There are notable benefits for the study of driver behaviour and vehicle feedback in adopting team-based models of SA. Even so, the notion that team SA relies on a combination of individual and shared awareness also reveals some fundamental limitations of the cognitive/experimental psychology approach to SA and its constituent require-ments. Whilst the notion of teams, in the loosest sense of the term, admits the possibility of greater degrees of emergence, the normative nature of SA still persists. In most team SA models 'good SA' is often still judged against some kind of normative benchmark or 'objective' referent. This runs counter to the potentially emergent nature of SA which is 'formative' rather than 'normative'. As before, this is not to suggest such models are not effective in certain practical situations.

It is important to again reiterate that in applying the most popular team SA approaches certain assumptions are being made. These are:

- It is still a cognitive phenomenon residing in the heads of individual human operators but with greater recognition of the 'ambient environment' and non-human agents in the system.
- There is a ground-truth of a situation available to be known but at the same time team SA may not be the simple summation of individual SA.
- Good SA can be derived with reference to expert or normative performance standards, although once again the emergent nature of team SA can affect the conclusions.

Numerous methods exist to measure facets of team SA but there is no formal and explicit method of team SA requirements capture. Being based to a large extent on individual models of SA the GDTA approach given by Endsley (1995) can, and often is, employed. Taking the more complex models of team SA the notion of a requirement becomes somewhat different to the simple presence or absence of an information object in the consciousness of an individual driver. The requirement becomes more formative and much more oriented around the outcomes to be achieved rather than the specific knowledge objects required to achieve it.

4.2.3 The System SA Lens

For a concept which began in the world of cognitive psychology, SA is no different to the Ergonomics discipline at large. The paradigm is shifting, however, and the strategic direction of the discipline is beginning to acknowledge increasing extents of systems thinking (e.g. Dul et al., 2012; Walker et al., 2010; Walker, 2016). A distinct part of the 'ergonomic offer' to stakeholders is its role within systems, and SA is naturally an important part of this (Stanton et al., 2017). That said, in sliding the systems lens across the SA frame some distinct challenges are revealed.

When one zooms out to consider the road transport system as a whole it is very evident it is not as deterministic as some SA theories or methods that might be applied to it. It does not behave like a machine with vehicles running on predefined routes according to a strict timetable. The people in the system are not 'rational optimisers'. Being in possession of a certain quantity of information does not guarantee predictable outcomes. The system is, instead, highly dynamic, often uncertain in its outcomes and behaviours, with multiple component parts, both human and technical. There exist comparatively broad constraints within which a wide variety of behaviours occur and from which strongly emergent phenomena arise. Indeed, it is precisely these strongly emergent phenomena which are becoming the focus of interest. This is because low-level individual driver behaviours can combine in interesting non-linear ways to give rise to powerful effects on things like emissions, fuel use, congestion, journey times and so forth, as we will see in Chapter 9. For now it is important to dwell on a potential mismatch, a mismatch between individual, and to some extent team, SA theories and methods founded on deterministic assumptions, and the highly systemic road transport system to which they are frequently applied.

SA was first discussed at a systems level by Artman and Garbis (1998), who called for a systems perspective to be adopted. It is also a theme picked up by Shu and Furuta (2005) in their attempts to enhance individualistic three-level models of SA using cooperative activity theory (Bratman, 1992). Stanton et al.'s (2006) Distributed SA (DSA) model is perhaps the closest the discipline currently has to a systems view of SA, to the extent the 'system view' in question is most firmly couched in systems theory. It is underpinned by three theoretical concepts:

1. Schema theory (e.g. Bartlett, 1932);
2. Neisser's (1976) perceptual cycle model of cognition; and
3. Hutchin's (1995b) distributed cognition approach.

In this model SA is viewed as an emergent property of collaborative systems. According to Stanton et al. (2006, 2009a, b) a system's awareness comprises a network of information on which different components of the system have distinct views and ownership. The model is scalable, with the network of information able to provide perspectives on individual, team and systems SA.

Scaling the model down to individual team members brings Neisser's perceptual cycle, as it was originally envisaged, into clearer focus. DSA can show where in the cycle individuals are and what is happening as they traverse it (Neisser, 1976; Plant and Stanton, 2014, 2015). At the individual and team levels, pre-existing 'geno-type' schema which people bring to a situation are combined with context-specific 'phenotype' schema (Plant and Stanton, 2014, 2015). This combination allows driver behaviour appropriate to a given context to be emitted. The idea of schemas calls into question the notion of shared SA (e.g. Endsley and Robertson, 2000). Rather than possess shared SA, which suggests team members under-stand a situation in the same way, the DSA model instead suggests team members possess unique, but compatible, types of awareness. Chapter 7 presents a good example of this by comparing car driver and motorcyclist SA. Both road users are participating in ostensibly the same situation but will naturally experience it in different ways. Instead of shared (i.e. identical) awareness there is instead compat-ible (and indeed incompatible) awareness. DSA argues this is what holds distributed systems like road transport together (Stanton et al., 2006, 2009a, b; Stanton, 2014) or, indeed, might split it apart. Each driver has their own awareness related to the goals they are working toward. This is, out of necessity, not the same as other drivers. They will have different goals, experiences, be driving different vehicles etc., but the awareness is compatible to the extent it enables them to co-exist in a common situation.

The compatible SA view does not discount the sharing of informa-tion, nor does it discount the notion different team members have access to the same information; this is where the concept of SA 'transactions' applies (Sorensen and Stanton, 2015). Transactive SA describes the idea that DSA is acquired and maintained through transactions in awareness arising from communications and sharing of information. A transaction in this case represents an exchange of SA between one agent and another (where 'agent' refers to humans and artefacts). As agents receive information it is integrated with other information and acted on, and then passed onto other agents. The interpretation placed on that information often changes as it passes hands. Transactive SA elements can form an interacting part of another actor's SA with-out any necessary requirement for parity of meaning or purpose. It is this transformation of situational elements as they cross the boundary from one system agent to another which bestows upon system SA an emergent behaviour (Figure 4.3). Flowing from this theory, therefore,

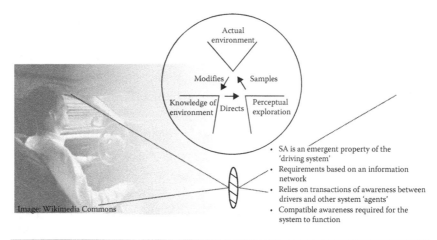

Figure 4.3 SA requirements projected through the systems SA lens.

are a radically revised set of assumptions or tenets (Stanton et al., 2006, 2017; Stanton, 2016):

- Situation awareness is an emergent property of sociotechnical systems. Accordingly, the system represents the unit of analysis rather than the individual agents working within it.
- Situation awareness is distributed across the human and non-human agents working within the system.
- Systems have a dynamic network of information upon which different operators have their own unique view and contribution, somewhat akin to a "hive mind" (Seeley et al., 2012). The compatibility between these views is critical to support safe and efficient performance, with incompatibilities creating threats to performance, safety and resilience.
- Systemic SA is maintained via transactions in awareness between agents. These exchanges in awareness can be human to human, human to artefact, and/or artefact to artefact, and serve to maintain, expand, or degrade the network underpinning the awareness within it.
- Compatible SA is required for systems to function effectively: rather than have shared awareness, agents have their own unique view on the situation which connect together to form systemic SA.
- Genotype and phenotype schema play a key role in both transactions and compatibility of SA.

The very notion of SA requirements becomes somewhat tenuous in the case of distributed SA. The idea of a 'requirement' does not align particularly well with the (non-deterministic) formative systems view. A 'requirement' conjures up ideas about there being 'a necessary condition' or a need to adhere to certain 'fixed' criteria. These sound normative rather than formative. But, to the extent that requirements are still useful, DSA does lend itself to their identification. In this case it is the identification of key information objects and their ownership. In identifying such nodes in the network it is then possible to ask how best to support SA transactions that might involve them. The foundation of a SA requirements analysis under the rubric of DSA is an information network (as deployed in Chapters 6, 7 and 8) rather than a task analysis. That being said, the former is an extremely useful tool in developing versions of the latter.

The systems approach certainly represents quite a dramatic shift in the SA paradigm. It is perhaps not surprising, then, that recent SA debates in the literature have tended to be somewhat adversarial (Endsley, 2015; Salmon et al., 2015; Stanton et al., 2015, 2017). This can be useful to drive out valid points of issue and subject theories and concepts to a stress test. It also makes the normally dry academic debate more interesting to read! Beyond all this, however, is a more measured picture. Simply put there is no one best theory; rather, it depends on the fundamental nature of the problem to which different SA approaches are being applied (Stanton et al., 2010). All have a role to play. If the practical SA issue being examined can be reasonably characterised as stable, relying on deviations from accepted normative practices, and focusing on individual drivers, then there are SA theories which match this question perfectly and will deliver the insights needed (e.g. Chapter 5). If, on the other hand, the problem can be characterised by a sociotechnical system in which SA is neither normative nor stable, and resides as a systems phenomenon rather than an individual one, then likewise, other approaches matched to these features will deliver the needed insights (e.g. Chapter 9). An attempt to summarise this diagrammatically is shown in Figure 4.4.

Returning to the beginning, the question to be answered in this chapter is a simple one: what are the driver's SA needs? We have established that SA is important and that vehicle feedback plays a role in its acquisition; now we need to understand what it might consist of.

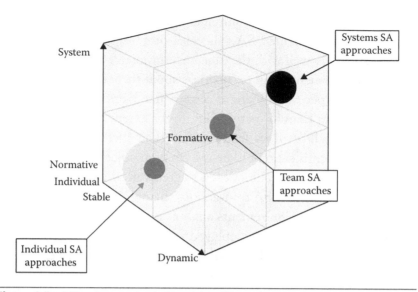

Figure 4.4 Different approaches to SA match to different features of ergonomic problems.

In doing this we can also confront a number of widespread assumptions in an interesting and systematic way. McRuer et al. (1977), for example, has stated that driving is comprised of a three-level hierarchy of navigation, manoeuvring and control, and these features have often persisted in the literature ever since. But is this really the case? Likewise, another well-worn statement which appears frequently in the literature (e.g. Reuben et al., 1988 etc.) is one originally made by Sivak (1996) which states driving is a 90% visual task. One has to ask, if this is the case why does the visual modality continue to be loaded with additional, and ever larger, displays and interfaces? One also has to ask whether it is even correct. Booher notes that with some forms of correction even severely visually impaired individuals can drive safely (1978), while Sivak himself was very keen to caveat the original '90% visual' statement:

> *"not only do we lack data from which to derive an accurate numerical estimate, but we lack a measurement system within which any numerical estimate would be meaningful. [...] The proliferation of such claims in the absence of direct evidence is a reminder that researchers should be careful about assuring the validity of the claims they are passing on"* (1996, p. 1081).

In being exhaustive and systematic the SA requirements analysis presented below enables us to critique *all* the sources of information and modes of feedback needed by the driver for the successful execution of the driving task, vision included of course. The findings also allow us to explore whether, and to what extent, those needs are being met.

4.3 What Was Done

4.3.1 Design

Given the nature of the question being asked, a 'task-based' SA requirements analysis, focusing on driver SA needs, is embarked upon. The requirements analysis deploys an existing and very comprehensive HTA of Driving (HTAoD) comprised of over 1600 bottom-level tasks and 400 plans. The following sections describe the theoretical rationale of HTA, the sources used to construct it, and the steps taken to validate it. The HTAoD is then partnered with an SA requirements template allowing each goal/sub-goal and operation to be assessed for the source of required information (individual, car and road environment) and feedback modality (visual, auditory and tactile). The task-based approach to SA requirements extraction is focused on the individual driver, but the coding scheme permits further insight into team and systems SA issues by examining where in the system information and feedback is coming from.

4.3.2 Methodology

4.3.2.1 Task Analysis   A task is defined as an "ordered sequence of control operations that leads towards some goal" where "a goal is some system end state sought by the driver" (Farber, 1999, p. 14/18). One of the ultimate, if simplistic, goals of driving is to depart from an origin and at some point reach a destination. This goal is, of course, comprised of many other sub-goals, such as reaching a destination quickly, comfortably, even enjoyably, and in such a manner as to avoid crashes. In order to fulfil the high-level goals of driving a vast array of individual tasks have to be performed more or less competently by the driver.

Task analyses of specific sub-sections of the driving task have been undertaken previously (Allen et al., 1971; Michon, 1985; Summala, 1996;

Gkikas, 2012) but it is interesting to note how the work of McKnight and Adams, dating from the 1970s, continues to be cited. We suspect the reasons for this are the daunting fine scale complexities of driving. Despite it being an 'everyday task' its fine-scale complexity surely deters many efforts to describe it exhaustively via task analysis. McKnight and Adams' work was prepared over 40 years ago for the U.S. Department of Transportation in order to "identify a set of driver performances that might be employed as terminal objectives in the development of driver education courses" (McKnight and Adams, p. vii). Although the study was initiated mainly from the point of view of driver tuition, it nonetheless provides some useful insights into the range of activities a driver has to perform. This range is extensive, with 43 primary tasks comprised of no fewer than 1700 sub-tasks. Despite providing these extremely useful insights into the range and quantity of tasks enacted by drivers, the stated purpose of McKnight and Adams' (1970) study limits its research applicability. So too does the fact it is not a Hierarchical Task Analysis (HTA). Neither is the report widely available.

4.3.2.2 Hierarchical Task Analysis Hierarchical Task Analysis (HTA) is a core ergonomics approach with a pedigree of over 40 years of continuous use. In the original paper laying out the approach, Annett et al. (1971) make it clear the methodology is based on an explicit theory of human performance. They proposed three questions as a test for any task analysis method, namely: does it lead to positive recommendations, does it apply to more than a limited range of tasks, and does it have any theoretical justification? Perhaps part of the reason for HTA's longevity is the answer to each of these questions is positive. To paraphrase Annett et al., the theory is based on goal-directed behaviour comprising a sub-goal hierarchy linked by plans (Stanton, 2006). Thus, performance towards a goal (such as driving a car to a destination) can be described at multiple levels of analysis. The plans determine the conditions under which any sub-goals are triggered. The three main principles governing HTA were stated as follows:

1. "At the highest level we choose to consider a task as consisting of an operation and the operation is defined in terms of its goal. The goal implies the objective of the system in some real terms of production units, quality or other criteria.

2. The operation can be broken down into sub-operations each defined by a sub-goal again measured in real terms by its contribution to overall system output or goal, and therefore measurable in terms of performance standards and criteria.

3. The important relationship between operations and sub-operations is really one of inclusion; it is a hierarchical relationship. Although tasks are often proceduralised, that is the sub-goals have to be attained in a sequence, this is by no means always the case." (Annett et al., 1971, p. 4)

It is important to fully digest these three principles, which have remained unwavering throughout the past 40 years of HTA.

In the first principle, HTA is proposed as a means of describing a system in terms of its goals. Goals are expressed in relation to objective criteria. The important point here is that HTA is a goal-based analysis of a system and that a 'system analysis' is presented by an HTA. These points can escape analysts who think they are only describing tasks carried out by people. In fact, by its very 'goal oriented' nature, HTA is quite capable of producing a systems analysis, describing both team work and non-human tasks performed by automation.

In the second principle, HTA is proposed as a means of breaking down sub-operations in a hierarchy. The sub-operations are described in terms of sub-goals. HTA is a description of a sub-goal hierarchy and the sub-goals can be described in terms of measurable performance criteria. To reiterate: HTA describes goals for tasks, such that each task is described in terms of its goals. 'Hierarchical Sub-Goal Analysis of Tasks' might actually be a better description of what HTA actually does, hence why it can replace Goal-Directed Task Analysis (GDTA) in SA requirements analyses.

The final principle states there is a hierarchical relationship between goals and sub-goals, and rules to guide the sequence in which sub-goals are attained. This means in order to satisfy the goal in the hierarchy its immediate sub-goals have to be satisfied, and so on. The sequence with which each sub-goal is attained is guided by the rules that govern the relationship between the immediate superordinate goal and its subordinates. These rules are expressed as 'plans' within the analysis. These principles can be seen in action within Figure 4.5.

In their original paper, Annett et al. (1971) present some industrial examples of HTA. The procedure described in the worked examples shows how the analyst proceeds in a process of continual reiteration and refinement. To start with, the goals are described in rough terms to produce an outline of the hierarchy. Progressive re-description of the sub-goal hierarchy could go on indefinitely, and Annett et al. (1971) caution that knowing when to stop the analysis is "one of the most difficult features of task analysis" (Annett et al., 1971, p. 6). The criterion often used is based on the probability of failure (P) multiplied by the cost of failure (C), known as the $P \times C$ *rule*. Annett et al. (1971) admit that it is not always easy to estimate these values and urges analysts not to pursue re-description unless it is absolutely necessary to the overall purpose of the analysis.

The enduring popularity of HTA can probably be put down to two main points. First, it is inherently flexible: the approach can be used to describe any system. Astley and Stammers (1987) point out that over the decades since its inception, HTA has been used to describe each new generation of technological system and its inherent abilities show no signs of weakness with future systems either. Second, it can be used for many ends: from personnel specification, to training requirements, to error prediction, to team performance assessment, system design and, in the present case, for defining situation awareness requirements.

4.3.2.3 The Hierarchical Task Analysis of Driving (HTAoD) In the present chapter we are interested in determining the structure of the driving task (is it really navigation, manoeuvring and control?), gaining a measure of the range of tasks performed by drivers (are they 90% visual?), and offering the wider research community a resource that has been 'developed once' but can be 're-used many times'. A comprehensive HTA of driving not only fills the knowledge gap surrounding exactly what it is drivers do, but it can also be used for a very large range of automotive design and engineering purposes. These include predicting driving errors with new in-vehicle technology, allocating driving functions between humans and technologies, road mapping the impacts of predicted technologies, and

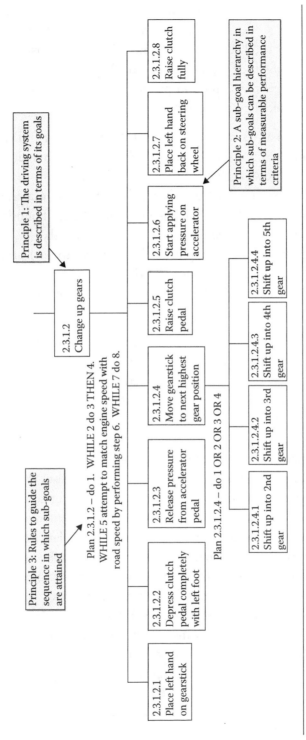

Figure 4.5   The principles of HTA have endured since the method's inception in the 1970s, giving rise to task analyses of this form.

much more. Indeed, in the region of 65% of all human factors methods, covering all the above domains and more, rely on task analysis as an input (Walker et al., 2010). The complete Hierarchical Task Analysis of Driving (HTAoD) itself is presented in the Appendix. It comprises over 1600 bottom-level tasks and 400 plans and is based on the following documents, materials and research:

- The task analysis conducted in 1970 by McKnight and Adams
- The latest edition of the UK Highway Code (based on the Road Traffic Act, 1991)
- UK Driving Standards Agency information and materials
- Mares et al.'s (2013) *Roadcraft* (the Police Foundation/Institute of Advanced Motorists drivers' manual)
- Subject matter expert input (such as police drivers) originating from our research in advanced driver training
- Numerous on-road observation studies involving a broad cross-section of drivers

The HTAoD begins by defining the driving activity (drive a car), setting the conditions in which this activity will take place (a modern, average-sized, front-wheel drive vehicle, equipped with a fuel-injected engine, being used on a British public road), and the performance criteria to be met (drive in compliance with the UK Highway Code and the Police Drivers' System of Car Control (Mares et al., 2013)). This exercise is extremely important in constraining what is already a large and complex analysis within reasonable boundaries. It also emphasises the point that the HTA represents a 'normative' description of driving task enactment, a 'specification' for good driving based on a robust foundation of experiential and training background material. It is a template against which real driving can be compared; alternative forms of vehicle technology analytically prototyped; deviations from normative standards revealed; and, of course, SA requirements for task success defined.

In the HTAoD the highest-level task goal is completely specified by six first-level sub goals, which altogether are completely specified by 1600 further individual operations and tasks. All of these are bound together by 400 plans of a logical form. These plans define how task enactment should proceed, often contingent upon specific conditions being met or criteria being present. The top level of the HTAoD hierarchy is shown in Figure 4.6 as a summary.

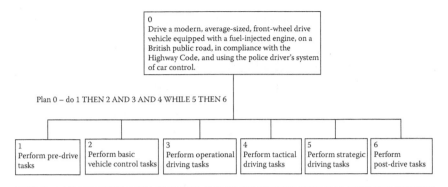

Figure 4.6 Top level of the HTAoD.

It is well known that HTA possesses a certain craft skill element and is contingent on the skills and impartiality of the analyst (Annett and Stanton, 1998, 2000). Although the prior task analysis of McKnight and Adams (1970) was referenced extensively it was not used as the structural basis of this new analysis. Despite this, the current HTAoD fell readily into similar categories of behaviour, suggesting a degree of concordance and reliability. In the HTAoD these categories are super-ordinate goals which comprise the following sub-goals:

Basic vehicle control tasks comprise the following:

- Task/Goal 2.1   pulling away from standstill
- Task/Goal 2.2   performing steering actions
- Task/Goal 2.3   controlling vehicle speed
- Task/Goal 2.4   decreasing vehicle speed
- Task/Goal 2.5   undertaking directional control
- Task/Goal 2.6   negotiating bends
- Task/Goal 2.7   negotiating gradients
- Task/Goal 2.8   reversing the vehicle

Operational driving tasks comprise the following:

- Task/Goal 3.1   emerging into traffic from side road
- Task/Goal 3.2   following other vehicles
- Task/Goal 3.3   overtaking other moving vehicles
- Task/Goal 3.4   approaching junctions
- Task/Goal 3.5   dealing with junctions
- Task/Goal 3.6   dealing with crossings
- Task/Goal 3.7   leaving junctions or crossings

Tactical driving tasks comprise the following:

- Task/Goal 4.1   dealing with different road types/classifications
- Task/Goal 4.2   dealing with roadway-related hazards
- Task/Goal 4.3   reacting to other traffic
- Task/Goal 4.4   performing emergency manoeuvres

Strategic driving tasks comprise the following:

- Task/Goal 5.1   perform surveillance
- Task/Goal 5.2   perform navigation
- Task/Goal 5.3   comply with rules
- Task/Goal 5.4   respond to environmental conditions
- Task/Goal 5.5   perform IAM system of car control
- Task/Goal 5.6   exhibit vehicle/mechanical sympathy
- Task/Goal 5.7   exhibit appropriate driver attitude/deportment

Below these higher-level goals are 1600 individual tasks and operations which, together, make up the total driving task. Obviously, 1600 bottom-level tasks, not to mention the collection of higher-level goals that bring us to this level of decomposition, cannot be presented adequately in the main body of text. The full HTAoD is therefore reproduced in full in the Appendix as a resource for other researchers. Looking at the above, however, the notion driving is a three-level hierarchy of navigation, manoeuvring and control (e.g. McRuer et al., 1977), whilst not necessarily incorrect, is certainly a simplification.

4.3.2.4 Validation of the Task Analysis   Two independent people are likely to approach the same task analysis in slightly different ways so the issue of reliability needs to be considered carefully (Stanton and Young, 1999). In most cases, if the analyst is given certain guidance on effective performance, then the definition of the user's goals will be quite straightforward and the higher-level goals quite robust. This is a key point. While there may be considerable debate about the nature and structure of individual tasks, there is usually less argument about the necessary goals the user is working towards, hence the focus on 'goal-directed' task analyses in SA requirements generation. Another point regards the ability of HTA to capture the full range and extent of the environment. What is often not appreciated is the relevant behaviour of the environment is largely specified by the plans. The plans specify, in effect, what environmental or contextual information needs to be responded to using what particular sequence of tasks. As such, a comprehensive task analysis will cover the larger part of the task context quite well which, in turn, helps to further aid the definition of SA requirements.

Validity refers to how well the method predicts what it sets out to predict, and a two-step process is employed to meet this aim. The first step is to observe or otherwise elicit real-world information that can substantiate the nature and structure of the task analysis. This is established by going back to the driver population whose task behaviour is being modelled. Depending on the needs of the analysis it is not always sufficient to base such validity checks on observation alone. A number of further structured ergonomic methodologies can also be used. Methods such as interviews and concurrent verbal protocols (among others) help to answer a key issue in driving research: that drivers may be performing the task in a highly skilled and largely automatic manner, unaware of the goals and plans underpinning their observable behaviour. Structured HF methods help to provide a robust means of uncovering such information. Several such methods, notably concurrent verbal protocols (e.g. Chapter 2), were relied upon to help construct the present HTAoD.

The second step is to validate the HTA through design, directly testing actual performance against HTA predictions. This is achieved by modifying aspects of the task that have been measured in the analysis, and systematically observing their effect on performance. The HTAoD has proved its worth in just such situations, including the work presented in this chapter and the next.

4.3.3 SA Requirements

Having constructed the HTAoD it then becomes possible to systematically work through all the identified operations, ascribing to them different sources and modalities of SA. These define the SA requirements of the driving task.

For the purposes of this analysis the pre- and post-driving tasks described in the HTAoD are excluded. This is because the vehicle is stationary and there is no interaction with the road or other traffic. This leaves 1100 bottom-level tasks or operations obtained from four parent/superordinate goals. Every one of these 1100 bottom-level tasks was coded using a mutually inclusive theme-based coding scheme. A theme-based analysis involves coding the meaning of task descriptions into shorter thematic units or segments (Weber, 1990). The coding scheme had two main categories. The coding scheme had two main categories. These were 'source of information' and 'feedback modality'. Source of information comprised: 'individual', 'car' and 'road environment'. Feedback modality comprised: 'visual', 'auditory' and 'tactile'. This categorisation follows that of Michon (1993).

The coding results were subject to inter- and intra-rater reliability (IRR) analysis using two independent raters. The independent raters made use of the same categorisation scheme as the main analyst. Across the six coding categories the inter-rater reliability was established at rho = 0.943 for IRR1 and rho = 0.921 for IRR2, both these values being significantly correlated ($p > 0.05$ for IRR1 and IRR2, $n = 6$). Having established satisfactory inter-rater reliability the results of the SA requirements analysis can now be presented.

4.4 What Was Found

Out of the 1100 driving tasks analysed 284 are basic, fundamental vehicle control tasks (e.g. controlling the vehicle in terms of speed and trajectory); 270 are operational driving tasks (e.g. operations for dealing with specific recurring situations such as junctions and crossings); 280 are tactical driving tasks (covering tactics for adapting to different road types and other traffic); and 266 are strategic driving tasks (dealing with rule compliance, navigation and other high-level strategies). It can be noted with interest just how much of the HTAoD refers to the development – tacitly – of driver SA. Based on this analysis it is clearly a major part of driving. It also tells us that while a complete lack of vision would render the driving task impossible to complete, there are far more non-visual cues than might be expected. Figure 4.7 shows this by providing a summary of the results for the modality of feedback (vision, auditory, tactile) while Figure 4.8 provides a summary of the results for source (car, other traffic, road).

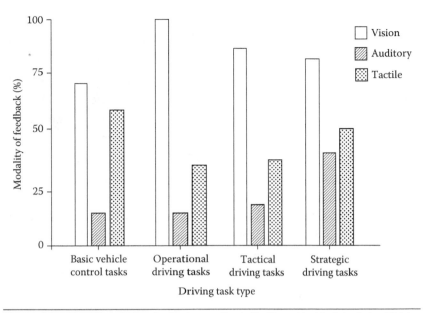

Figure 4.7 Modalities of driver feedback according to the Hierarchical Task Analysis of Driving (HTAoD).

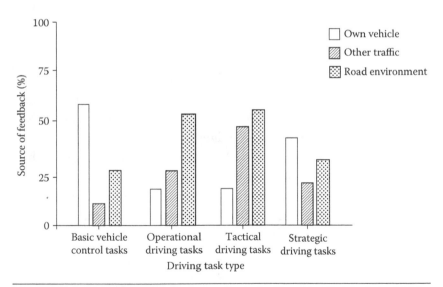

Figure 4.8 Sources of driver feedback according to the Hierarchical Task Analysis of Driving (HTAoD).

Figure 4.7 communicates something quite powerful. It provides a direct answer to the question of whether driving really is a '90% visual task'. Ignoring for a moment the desirability of making such a numeric estimate – which we are equally cautious about – we can answer that, no, it is not 90% visual. According to this analysis it is 85.18% visual. But it is also 22.26% auditory and a surprising 46.03% tactile. Vision clearly dominates, as one would expect, but there is a strong multi-modality which, based on some of the current literature, one would not expect. Great care needs to be exercised in providing numerical estimates like these and we agree strongly with the sentiments expressed earlier that "researchers need to be careful about assuring the validity of the claims they are passing on" (Sivak, 1996, p. 1081). What we can say is that SA requirements based on HTA provide 'a form of' data from which to provide a numerical estimate of this sort. Whether or not a numerical estimate is particularly useful or helpful is a moot point. More interesting to us is the analysis shows very clearly that driving is multi-modal.

It will be recalled that the coding scheme used to analyse each of the individual tasks is 'non-mutually exclusive'. This means the bars in Figure 4.7 do not total to 100%. Individual driving tasks can – and are – coded to more than one feedback modality where it is appropriate to do so. Table 4.1 breaks down the extent of multi-modality for each level of driving task where it can be observed that, apart from strategic driving tasks, there are over half as many more feedback modalities as there are driving tasks. A 'headline grabbing' reading of the wider results is indeed that driving is visually dominated but according to Table 4.1 this would be to miss a great opportunity. Not only does the analysis make no judgement about the relative importance of a feedback mode, it is also clear just how important other, non-visual feedback modalities are.

Looking in detail at Figure 4.7 we can, for example, observe just how significant tactile feedback is for basic vehicle control tasks: it

Table 4.1 Driving Tasks within the Task Analysis Can Be Coded with More Than One Feedback Modality

DRIVING TASK LEVEL	EXTENT OF MULTI-MODALITY (%)
Basic driving tasks	52.4
Operational driving tasks	49.4
Tactical driving tasks	41.8
Strategic driving tasks	70.3
Mean	53.5

rivals visual feedback in quantity. Operational and tactical driving tasks see the role of tactile feedback falling back, but over 30% of such tasks still require it. Likewise for auditory feedback, between 16% and 19% of operational and tactical tasks require an auditory component. Strategic driving tasks, as noted in Table 4.1, require even more multi-modality, with such tasks founded on 84% visual feedback, 37% auditory and 50% tactile. It is important to keep restating that the number of times a feedback modality is coded in the analysis says little about its relative importance. Despite this, the multi-modality of driving is clear and an analysis like this provides a powerful tool for thinking and reasoning about future car designs. Are such designs providing adequate multi-modal support for the driver tasks that need supporting? In which direction are feedback modalities heading (less tactile and auditory feedback perhaps)? And what happens as ever larger chunks of the driving task are transferred to automation: what are the driver's multi-modal feedback requirements then? This is where a systemic view of SA can be helpful.

When dealing with SA transactions not only does a systematic task analysis help to identify the nodes in the information network, but also the means by which some of them will be transacted – visual, auditory and/or tactile. Stretching out into the realm of future research is the question as to whether changes in the inherent multi-modality of driving will affect the transactive nature of driver SA. For example, an 'irreversible' steer-by-wire system (see Chapter 3) removes a form of tactile feedback through which information elements are transacted; so how else could those elements be transacted? How does their absence now impact on other parts of the distributed SA network? These are important questions.

Where does the feedback come from? Certainly in the case of basic driving tasks – those most likely to become automated in future – it comes from the car. More than 80% of the information requirements needed for these low-level control tasks stem from the feel of the controls, the weight of the steering, the noise of the engine, and other inherent vehicle feedback properties. As we move up the task hierarchy to consider operational driving tasks, such as overtaking, passing, negotiating junctions and so forth, the picture begins to change. SA requirements seem to be derived much more from the road environment (45%), but again, the car is providing a meaningful contribution (32%). Indeed, moving further still through the task levels to consider tactical and strategic tasks the source of driver feedback becomes increasingly road-based (between 54% and 56%) while the role of the car shrinks relatively (to approximately 18%). This again is interesting when one considers recent trends in vehicle-to-vehicle and vehicle-to-infrastructure communications. How could all of this be re-imagined so as to support driver SA requirements in new ways? Could this new technology provide even better ways for driver SA requirements to be extracted from the road and presented to drivers? The possibilities are tantalizing. Interestingly, when considering strategic driving tasks such as navigation, surveillance and defensive driving, the role of the car in meeting SA requirements asserts itself once more. 43% of driver feedback required for these high-level driving tasks stems from the vehicle, with 23% derived from the actions/behaviour of other traffic, and 33% from the road environment.

So where does the feedback come from? The answer – bearing in mind the analysis makes no judgement about the relative importance of feedback sources – is 43% of it comes from the wider road environment and context; 35% of it comes from the vehicle itself; and 26% comes from the actions and behaviour of other traffic. What is interesting is that, unlike the multi-modality of feedback type, feedback source is more-or-less 'single point'. SA requirements might take the form of auditory, tactile and/or visual modes but those modes tend to be derived from just one source, either the car, other traffic, or the wider road context. Only on few occasions is there multi-source driver feedback, and only then within tactical driving tasks where, presumably, the opportunity to gain feedback from multiple sources is present. This is an interesting feature which may be an artefact of the hierarchical form of the tasks analysis which, by its nature, divides the task in a way that might favour a single source of feedback. The efficacy of this finding, and its relevance, requires more in-depth analysis. What is clear, and also much more within the remit of the current chapter, is just how important the vehicle is as a means of meeting driver SA requirements. It is also clear how important other traffic and the wider road environment are too.

4.5 Summary

Back in the 1978 Mini, engine spinning noisily at 4200 rpm at an indicated 70 mph, tyres thumping rhythmically over the expansion joints in the motorway, the observation that driving is much more than just a visual task is not hard to imagine. Likewise, subjecting a 1978 Mini to an SA requirements analysis and then comparing it to a 2018 steer-by-wire Infiniti Q50 would be equally dramatic. An HTA-based SA requirements analysis, then, clearly tells us something.

Firstly, while it is possible to ascribe a value to the visual portion of the driving task, the headline is that driving is multi-modal. The acquisition of awareness relies on multiple overlapping cues derived from visual, auditory and tactile modes.

Secondly, SA requirements, like SA itself, can be projected through multiple theoretical lenses. The individual SA lens would use SA requirements as a form of 'ground truth' against which a target vehicle, design intervention, or other experimental manipulation can be tested using methods such as SACRI (Hogg et al., 1995) or SAGAT (Endsley, 1988). Indeed, this very same individual SA lens is slid across the frame and deployed in the next chapter to good effect. The team SA lens can also be used. By coding the sources of feedback into vehicle, other traffic and road environment categories, it is possible to explore the notion of information exchange. At what point, for example, is the driver exchanging information with other road users and the wider environment, and vice versa in the cyclical process described by Smith and Hancock (1995) and Salas et al. (1995)? It also allows notions of shared and mutual awareness to be explored: for each 'agent' in the situation, to what extent are their requirements being met and, indeed, are they identical?

For us, though, the systems lens is perhaps the most interesting. From the present analysis it is clear just how manifest the transactive nature of distributed SA is. There are multiple modalities of feedback derived from multiple sources, and they all connect within a network of information objects. We erode those connections, or unwittingly create new ones, at our peril. We might be developing painful noise-induced headaches inside our 1978 Mini but let's imagine the polar opposite: an extreme situation of no driver feedback, literally zero. The vehicle as a kind of isolation tank or mechanical, indeed, 'existential' bubble. A case whereby all the non-visual SA requirements are dialled completely out. What would that be like? The next chapter takes a look.

5

THE IRONIES OF
VEHICLE FEEDBACK

5.1 Introduction

5.1.1 Ironies (and Problems)

Lisanne Bainbridge, in her seminal paper, defines irony as a: "combination of circumstances, the result of which is the direct opposite of what might be expected" (1983, p. 775). Writing of the 'ironies of automation' in industrial process control contexts, she identifies the main ones as thus:

> *"the increased interest in human factors among [automotive] engineers reflects the irony that the more advanced a [vehicle automation] system is, so the more crucial may be the contribution of the [driver]"* (p. 775).

> *"the irony that one is not by automating necessarily removing the difficulties, and also the possibility that resolving them will require even greater technological ingenuity than does classic automation"* (p. 778).

The main 'problem of automation', as Donald Norman goes on to describe in his equally seminal paper (1990), is that:

> *"automation is at an intermediate level of intelligence, powerful enough to take over control that used to be done by people, but not powerful enough to handle all abnormalities. Moreover, its level of intelligence is insufficient to provide the continual, appropriate feedback that occurs naturally among human operators"* (p. 585).

He goes on to propose that the main problem:

> *"is not the presence of automation, but rather its inappropriate design. The problem is that the operations under normal operating conditions are performed appropriately, but there is inadequate feedback and interaction with*

the humans who must control the overall conduct of the task. When the situations exceed the capabilities of the automatic equipment, then the inadequate feedback leads to difficulties for the human controllers" (p. 585).

The design risks are potentially significant. We risk designing out precisely the situation awareness (SA) requirements drivers want and need, in favour (perhaps) of focusing on yet more visually dominant information sources from arbitrary and disconnected forms of technology, unlearning lessons acquired from other domains as well as a rich automotive developmental history in doing so. Taken to its limit, the argument would run to a future vehicle that places drivers in a form of sensory, possibly even existential bubble, isolated like Zuboff's famous printing press operators (1988), immune to the continuous auditory and tactile feelings of system function, unable to keep track of critical variables, error prone in unexpected ways, and not able to successfully intervene even in cases where the automation requires it. A 'sensory bubble' of this sort is created in the experiment that follows with interesting results.

The corresponding design opportunity is as follows: if we need to rely on humans to assume control and perform the tasks left over from automation, then vehicle feedback represents a potentially underused means by which to do it. Instead of a bubble of isolation what about a Norman-esque 'chatty co-pilot', a car that behaves like another member of the team, a mechanical and electronic extension of the driver more like 'KITT' (from the TV serial *Knight Rider*) rather than the malevolent HAL (from the film *2001: A Space Odyssey*)? Having already used 15 years of vehicle design evolution as an independent variable in Chapter 2, and having explored the sources and modalities of vehicle feedback, and the driver's SA requirements therein (Chapter 4), it is time to progress further. This is achieved by examining how driver SA is represented internally when vehicle feedback is manipulated under laboratory conditions. This individualistic view of SA is not suitable for all the questions and ironies we wish to explore, but it is useful to try and understand the adaptive capacity of drivers, how their SA changes in response to changing combinations of vehicle feedback, and specifically what happens when we place them in a Zuboff-esque condition of complete and total vehicle feedback deprivation.

5.1.2 *Measuring Situation Awareness*

The fundamental problems and ironies of automation find clear analogies within the automotive sphere. It follows, therefore, that the methods applied previously in other domains to examine feedback and SA might also be appropriate. Two of these, SAGAT (Situation Awareness Global Assessment Technique) (Endsley, 1988) and SACRI (Situation Awareness Control Room Inventory) (Hogg et al., 1995) are both well-established SA measurement methods used widely in aviation and high-hazard sectors. At the time this study was performed, their use in the context of driver SA was (and to a large extent still is) novel. In crossing the divide between problem domains the opportunity was taken to revisit both methods and review them critically.

SAGAT and SACRI both utilise a probe recall method consistent with their focus on individual SA. This involves freezing an experimental scenario, be it a process control 'micro-world' or flight simulator, and then accurately probing a participant's situation awareness by asking questions about the current state of the situation, for example, "What position was xyz aircraft in?" or "What was the state of this instrument?". The closer the match between the participant's awareness of the situation, and the 'objectively' defined state of the situation, the better the person's state of SA is judged to be. The cause and effect logic of this approach is undeniably strong, but we would argue not strong enough to justify the claim the methods are an 'objective' means to measure SA (Endsley, 1988, 2000; Hogg et al., 1995).

If measurement is defined as the process of converting observations into quantities through theory, then objectivity is achieved when "within the range of objects for which the measuring instrument is intended, its function [is] independent of the object of measurement" (Thurstone, 1928, p. 547). Moreover, the results of the measurement are independent from the conditions in which it took place. Against these strict criteria it seems unlikely any method in behavioural science could properly be considered objective. An enhanced version of SAGAT/SACRI is required, if not to achieve true objective measurement, then at least to recognise and attempt to circumnavigate some of the more prominent issues.

A revised method requires set procedures to systematically assess the publicly observable and measurable state of the world that is to be compared with the participants' SA of that world. The aim, in other words, would be to place less reliance on experts judging the 'goodness' or 'badness' of a situation. This is a feature common in both SAGAT and SACRI.

The constructs being measured, and from which a response from the participant is required, can also be regarded as multi-dimensional. They involve the participant making a complex cognitive appraisal of the situation in order to reach certain judgements about it. Again, far from being objective, this is unavoidably subjective (Annett, 2002).

The SA requirements analysis presented in the previous chapter points to a more structured way of defining 'probes' that hold up better (but not perfectly) to the precepts of objectivity. While examining issues of SA measurement it is also important to note a further potential confound which afflicts both SAGAT and SACRI. Both tend to make use of parametric statistics to compare constructs measured along rating scales (Lodge, 1980). Rating scales are ordinal not continuous.

The end point of this brief methodological discussion is as follows. Despite the appeal inherent in a logic centred on the idea of a 'ground truth' or objectively knowable situation, and the extent to which an individual's performance against it compares, there are important subtleties. These will be developed much further as the monograph progresses. For now, though, it is possible to rely on Bell and Lyon (2000), who propose that "all aspects of momentary SA are eventually reducible to some form of [...] information in working memory" (p. 42). From this simple expedient emerges a strategy to mitigate the concerns above. This is to pursue a more atomistic level of analysis, tapping into this information in working memory, where discrete SA requirements support discrete driving tasks. This procedure avoids any complex cognitive appraisal of complex situation probes. What is being measured instead are discrete uni-dimensional aspects of an individual's SA. Under this paradigm a comparison based on information actually existing in the world and information actually perceived by the individual can be framed differently, as 'sensitivity', assessed using rating scales, and tested with non-parametric techniques.

Sensitivity is a publicly observable attribute measured in physical units; it is separate from the object of measurement, and largely independent from the conditions in which the measurement took place. Admittedly, a method based on these concepts still cannot be regarded as truly objective, but it does provide a highly systematic set of procedures and controls and a formal quantitative structure; and avoids some of the more objectionable measurement issues described above. Unfortunately, in addressing some of these issues with an atomistic, information-based approach to individual SA, the phenomenon is becoming quite narrow and less like the naturalistic phenomenon of driver SA as it exists 'in the real world'. This is why subsequent chapters project the concept of SA through different types of theoretical and methodological lens. For now, though, it is still sufficient to enable us to explore what happens to driver SA when vehicle feedback is manipulated in quite a dramatic fashion.

5.2 What Was Done

5.2.1 Design

The laboratory experiment was based on a within-subjects probed recall paradigm using a driving simulator and a predetermined virtual road course. The dependant measures were the participant's responses to two sets of rating scales completed during 36 pauses in the driving simulation. These pauses were placed at varying time intervals along the virtual road course, in identical locations for all participants. No prior warning was provided to participants that a pause was about to occur.

The first set of rating scales comprised seven items. They probed the participant's confidence level as to the presence or absence of SA probe events in the environment using a seven-point confidence rating scale. Consistent with the 'atomistic' information-based approach to the analysis is that probe items were derived from the SA requirements developed in the previous chapter.

The second rating scale was drawn from the SART questionnaire (Taylor et al., 1994). This required participants to directly rate their own situational awareness along a single 20-point scale from zero (a poor grasp of the situation) through to 20 (a complete picture of the situation). The two sets of rating scales were identical for all participants.

Table 5.1 Eight Levels of Simulated Vehicle Feedback

NUMBER OF FEEDBACK MODALITIES	CONDITION	NOTES
1 Feedback Modality	Condition 1	Visual feedback only (baseline)
2 Feedback Modalities	Condition 2	Visual + Auditory
	Condition 3	Visual + Steering feel
	Condition 4	Visual + Under seat resonators
3 Feedback Modalities	Condition 5	Visual + Auditory + Steering
	Condition 6	Visual + Auditory + Resonators
	Condition 7	Visual + Steering + Resonators
4 Feedback Modalities	Condition 8	Visual + Auditory + Steering + Resonators

The independent measure was the feedback provided by the simulation vehicle and this had eight levels, chosen at random following each pause, as shown in Table 5.1.

Signal Detection Theory (SDT) was used as the overarching analysis paradigm. Conceptually SDT provides a measure of participants' ability to discern a 'signal' from a background of 'noise'. Condition 1, wherein drivers have just visual feedback during the simulated driving task, serves as the 'noise' trial upon which the seven further conditions of vehicle feedback were overlain. The extent to which drivers can correctly discern SA probe items from this 'noise' background enables a measure of sensitivity to be derived. Sensitivity, in turn, serves as a proxy measure for individual SA performance. It should be noted that Condition 1, the visual-feedback-only baseline, is the equivalent of having all non-visual SA requirements removed. An extreme condition: a sensory driver SA bubble.

5.2.2 Participants

Thirty-five drivers took part in the experiment. This offered a power level of 0.8 (Faul and Erdfelder, 1992; Siegel and Castellan, 1988; Cohen, 1988). It means the study is powerful enough to detect medium (or larger) effects with an 80% probability. This is more than satisfactory given the effects which emerged in the opening Chapter under much less favourable experimental conditions.

Following the challenges identified previously in terms of recruiting a gender-balanced sample, strenuous efforts were made to address this here. As such, recruitment involved a large variety of means over a lengthy time period. These means included a national radio advertisement and local publicity via leaflets, newsletters and email campaigns. Despite all this, around three quarters of the participants were still male (77%). All age groups had some representation though, from 17 years through to 61–70 years. The modal age category was 21–25. All drivers held a valid driving licence and had at least one year of driving experience (over half had more than six years).

5.2.3 Materials and Methodology

5.2.3.1 Driving Simulator The driving simulator was constructed specifically for this study. It was based on a UK specification Ford Mondeo, with a high-resolution image measuring 3.5 by 2 metres projected onto a Perlux® screen in front of the car using a 3200 lumen projector. This provided a field of view of approximately 60 degrees horizontal and 35 degrees vertical, as illustrated in Figure 5.1 below.

The simulator was explicitly designed to retain the look and feel of a standard road vehicle as far as practical. The driver interacted with the vehicle's standard controls and the simulation computer manipulated the visual scene dependent upon the driver's control inputs. Apart from visual information the simulation also provided auditory feedback, tactile steering force feedback, and tactile feedback through

Figure 5.1   Typical visual scene within the driving simulation.

the driver's seat. These sources of feedback could be independently switched on or off remotely by the experimenter.

Auditory feedback was presented in Dolby Pro-Logic™ four-channel surround sound through the vehicle's original interior loudspeakers. An additional hidden centre speaker was also fitted, along with an external low-frequency subwoofer unit. Steering feel was enabled by a torque motor assembly incorporated within the original Ford Mondeo steering column, which was disconnected from the host vehicle's steering system. Reactive force feedback and self-centring force could be defined through four levels up to a maximum (at the steering wheel rim) of 1.25 Nm. A degree of whole body vibration was also provided by two electromechanical resonators fitted to the underneath of the driver's seat. These were fed with low pass filtered inputs from the audio signal and a 50-Hz sine wave fed from a signal generator.

5.2.3.2 Simulation Road Course The road course for the simulation was 24 miles long and designed to offer a balanced mix of different road types and environmental conditions. These included both rural and urban driving.

5.2.3.3 Objective States for the SA Probes At any given pause in the simulation the objective, publicly observable and measurable state of the simulation was known. The underlying logic is that a comparison of subjective confidence ratings provided by the participating drivers compared to the actually existing uni-dimensional objective information in the world provides the basis for a measure of sensitivity. In this case high sensitivity would imply the driver's subjective ratings of information present were in accordance with the true objective presence of information in the world, and vice versa. As noted earlier, the specific information in the world was selected based on an SA requirements analysis.

5.2.3.4 Probes for SA Subjective States Development of the probe items relied on the Hierarchical Task Analysis of Driving (HTAoD) described in Chapter 4. The HTAoD was employed to systematically and exhaustively elicit the SA requirements of normative driving performance. This analysis proceeded at the HTA 'Sub-Goal' level and not at the level of individual tasks and operations. Endsley and Garland (2000) argue that goals, as distinct from tasks, provide the proper basis for decision making in studies such as this. Other advantages are the SA requirements are taken out of a detailed context applicable to only a narrow domain, or even a specific vehicle, and they do not favour any particular feedback channel or modality. Response bias is thus avoided. The wording of the probe items was also referenced to what drivers actually did say about features and events in a naturalistic driving environment. This data was supplied from the on-road study described in Chapter 2. This enabled probe items to be couched in naturalistic terms and presented in a form of words that was recognisable to drivers.

5.2.4 Procedure

Participants were briefed at the start of the trial. The simulation would pause was explained, with the screens going blank; this was the cue to complete the probe item scales. Whilst driving, participants were instructed to remain close to 70 mph and on the left-hand side of the road. This was a reasonably demanding yet sensible and achievable speed for the simulated road conditions. The experimental phase was preceded by a practice drive.

During the experimental phase the simulation was paused at predetermined points on the course 36 times for a total driving time of approximately 45 minutes. During the pauses the driver completed the scales appropriate to the pause. At the same time the experimenter manipulated the feedback presentation (or appeared to manipulate it) prior to the simulation re-starting. The driver signalled to the experimenter when they were ready to commence.

5.2.5 Data Reduction

A Signal Detection Theory (SDT) paradigm was used. This is based on measuring the extent of a driver's sensitivity to the SA probe items. SDT requires a baseline feedback condition (in this case visual stimuli only) which is regarded conceptually as the 'noise' trial. The other feedback conditions are all superimposed onto the baseline and represent seven further and distinct 'signal' trials. Taking into account a few instances of missing responses, this gave a mean of 233 probe item questions completed by each driver during the simulation. The number of trials completed per condition are presented in Table 5.2 and totalled no fewer than $n = 8148$.

The objective state of the scenario would either be represented in the analysis as a one or a seven on the confidence rating scale (i.e. simply 'Yes', probed information was present, or 'No', probed information was not present). The driver's subjective confidence ratings were entered into the analysis directly against these scores and compared.

The responses provided by the drivers were organised into the following taxonomy. If the driver rated confidence that a stimuli was present and it was, this is a Hit (H). If the stimuli was rated with some degree of confidence it was not present, and objectively it was not, this is a Correct Rejection (CR). A Miss (M) is when confidence is expressed that a stimuli was not present and objectively it was, and a False Alarm (FA) is confidence being shown that a stimuli was present when in fact it was not. This taxonomy of Hits, Misses, False Alarms and Correct Rejections (H/M/FA/CR respectively: Table 5.3) forms the basis of SDT (Green and Swets, 1966).

Three decision criteria were also defined. A Conservative criterion would be to only accept 'very confident' scores into the analysis. A Medium criterion would permit anything above a 'reasonably confident' score. The Risky criterion would accept anything above a 'maybe/possibly confident' score.

The next phase of the analysis was to sort the data according to feedback condition, and sum the Hit, Miss, False Alarm and Correct

Table 5.2 Number of Signal and Noise Trials per Vehicle Feedback Condition ($N = 35$)

FEEDBACK CONDITION	FEEDBACK STIMULI	NUMBER OF TRIALS	SDT TRIAL TYPE
1	Visual	980	Noise
2	Vis + Auditory	1078	Signal + Noise
3	Vis + Steering	952	Signal + Noise
4	Vis + Resonators	1099	Signal + Noise
5	Vis + Steer + Aud	882	Signal + Noise
6	Vis + Res + Aud	1057	Signal + Noise
7	Vis + Res + Steer	1120	Signal + Noise
8	Vis + Aud + Res + Steer	980	Signal + Noise

Table 5.3 SDT Taxonomy

TRIAL TYPE	RESPONSE TYPE	
SIGNAL Present (probe item was present in the situation to be perceived)	HIT (driver responds with confidence that probe item was present)	MISS (driver responds with confidence that probe item was not present)
NOISE (probe item was not present in the situation to be perceived)	CORRECT REJECTION (driver responds with confidence that probe item was not present)	FALSE ALARM (driver responds with confidence that probe item was present)

Rejection scores for each condition and each decision criterion. The Hit and False Alarm rates were then converted into proportions of the actually existing signal/noise events, and a correction factor was applied to the data prior to computing the sensitivity measure d-prime (d') (Snodgrass and Corwin, 1988; MacMillan and Creelman, 1991; McNicol, 1972).

5.3 What Was Found

5.3.1 'Objective' Situation Awareness (Probed Recall)

5.3.1.1 Calculation of Sensitivity Measures Higher d' values are associated with more 'Hits', and therefore more accurate and confident ratings about the presence/absence of probed events, which in turn signals better driver SA. The results for overall median d' versus vehicle feedback condition are shown in Figures 5.2 through 5.4.

5.3.1.2 Statistical Tests on D-Prime versus Feedback Condition Non-parametric procedures were employed because the use of rating scales undermines the assumptions of parametric tests (e.g. Lodge, 1980) and a more conservative approach was favoured. Two further advantages were discovered. Firstly, non-parametric statistics enable the full data set to be used and avoid a number of significant problems in accounting for outliers and variability in the sample. Secondly, parametric tests are based on comparing the mean d' across individuals and across experimental conditions. This procedure actually places d' in error (Ingleby, 1968). Non-parametric tests use ranking methods and median d', therefore avoiding this problem.

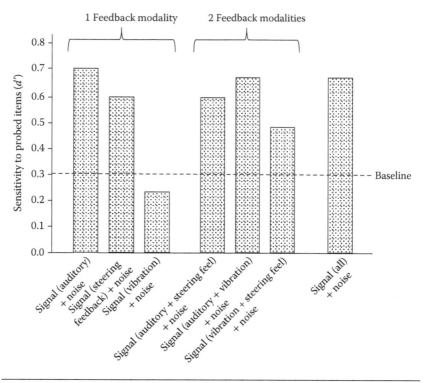

Figure 5.2 Values of d' for the conservative criterion versus feedback condition ($N = 35$).

The Friedman test reveals statistically significant differences between sensitivity (d') and vehicle feedback condition. The significant differences occurred within every response criterion group (Conservative criterion: $X_r^2 = 22.78$, df = 7, $p < 0.01$, Medium criterion: $X_r^2 = 26.23$, df = 7, $p < 0.01$, and Risky criterion: $X_r^2 = 23.13$, df = 7, $p < 0.01$). These findings support the impression developed in the previous Chapter. Driving is multi-modal and providing feedback in addition to vision increases individual driver SA.

5.3.1.3 Comparisons with the Visual Baseline The visual baseline represents the case of 'Zuboff isolation' described earlier. This is the example of driving in complete silence, with no steering or road feel: the feedback 'bubble'. Respondents were noted as saying how strange this felt. Whilst not formally measured in this study it was interesting to observe how steering performance and speed control seemed to suffer. Post-hoc testing against this vision-only

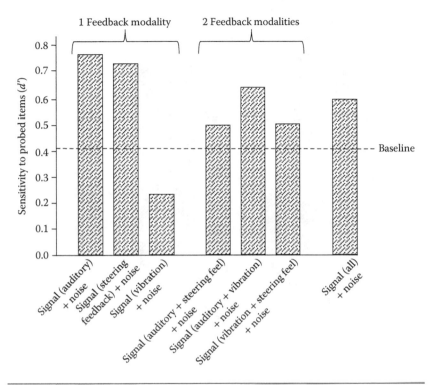

Figure 5.3   Values of d' for the medium criterion versus feedback condition ($N = 35$).

baseline condition for the Friedman procedure involves the following formula:

$$\left[\overline{R_1} - \overline{R_u}\right] \geq q(\alpha, \#c)\sqrt{\frac{k(k+1)}{6N}}$$

where:

R = baseline and other group pairs
q = critical value (see Siegel and Castellan, 1988, p. 321)
α = significance level
$\#c$ = number of comparisons
k = conditions
N = participants

This formula is used to derive a critical value (Siegel and Castellan, 1988).
The absolute difference between pairwise comparisons of the baseline
and every other condition is referenced against this critical value. The
critical value is determined to be 1.52 (for significant differences between

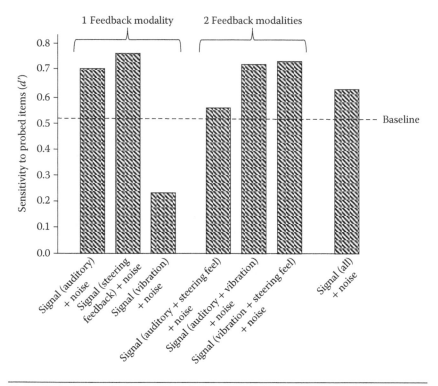

Figure 5.4 Values of d' for the risky criterion versus feedback condition ($N = 35$).

Table 5.4 Results of Baseline Comparisons between the Visual Baseline (noise) and the Remaining Signal + Noise Conditions

DIFFERENCE	CONSERVATIVE	MEDIUM	RISKY
Visual vs. Visual + Auditory	−2.02[a]	−1.64[a]	−0.99
Visual vs. Visual + Steering Feedback	−1.26	−1.42	−1.33
Visual vs. Visual + Resonators	−0.20	0.31	0.53
Visual vs. Vis + Steer + Aud	−1.23	−0.62	−0.26
Visual vs. Vis + Aud + Res	−1.62[a]	−1.89[a]	−1.79[a]
Visual vs. Vis + Steer + Res	−0.56	−0.82	−0.87
Visual vs. Vis + Aud + Steer + Res	−1.59[a]	−1.37	−0.90

[a] denotes significant difference at 5% level

conditions present at the 5% level) and the pairwise comparisons that exceed the critical value are highlighted in Table 5.4.

Table 5.4 shows that sensitivity towards SA probe stimuli increases when certain additional feedback modalities are presented on top of vision. In the strictest Conservative criterion, where drivers indicate a very high degree of confidence in their responses, auditory feedback, auditory feedback combined with tactile feedback (under-seat resonators), and all these modalities combined again with steering force feedback, significantly increase sensitivity towards the SA probe items compared to vision alone.

In the Medium criterion, once again auditory feedback is associated with higher sensitivity than vision alone, as is auditory feedback combined with tactile feedback. In the Risky decision criterion, only the combined effect of visual, auditory and tactile feedback yields increased sensitivity from the baseline. That is, in the Medium and Risky criterion points, where confidence ratings are moderate to low, steering force feedback appears not to make a significant contribution towards increasing driver SA on its own. It does, however, increase SA when it is combined with all other modalities in condition eight. Auditory feedback, on the other hand, appears to offer a consistent increase in driver SA above vision alone, regardless of confidence criterion. This reinforces a point made in the previous chapter. The point was made several times that coding SA requirements in terms of their modality does not say anything about their importance. In this case, the modality which earlier scored the lowest (i.e. auditory feedback) actually seems to be very important.

5.3.1.4 Multiple Comparisons Pairwise post-hoc testing for the Friedman procedure involves the following formula:

$$\left[\overline{R_u} - \overline{R_v}\right] \geq Z / k(k-1)\sqrt{\frac{Nk(k+1)}{6}}$$

where:

R = baseline and other group pairs
Z = Z value
k = conditions
N = participants

Table 5.5 highlights where significant ($p < 0.05$) pairwise comparisons between vehicle feedback conditions are reported.

Table 5.5 shows that amongst conditions which yielded a significant change in sensitivity from the visual baseline, none of them are

Table 5.5 Significant ($p > 0.05$) Results of Multiple Comparisons between All Feedback Conditions

DIFFERENCE	CONSERVATIVE	MEDIUM	RISKY
Vis + Res < Vis + Aud	−1.82[a]	−1.95[a]	−1.86[a]
Vis + Res < Vis + Aud + Res		−2.20[a]	−2.32[a]

[a] denotes significant difference at 5% level

significantly different from each other. In other words, sensitivity appears not to be significantly increased despite presenting two, three or even four feedback modalities. The most consistent finding is that auditory feedback yields a significant increase above the visual baseline condition on its own, and that every other condition which yields a significant increase in sensitivity includes it.

5.3.1.5 Do the Findings Matter? The actual effect size observed in the experiment can now be calculated in order to provide a numerical measure of association between vehicle feedback and individual driver SA. At the overall level of analysis the effect size is found for the Conservative criterion to be $R_{phi} = 0.81$, for the Medium criterion to be $R_{phi} = 0.87$, and for the Risky criterion to be $R_{phi} = 0.81$. These values are indicative of a strong positive association between the vehicle feedback presented in the driving simulator and consequent measured increases in sensitivity. The findings do matter. Anecdotally we had good reason to suppose that vehicle feedback impacted on SA. This was confirmed during on-road testing and is confirmed again under more controlled circumstances.

5.3.2 'Subjective' Situation Awareness (Self-Report)

In direct contrast to rating the objective presence or absence of SA requirements via the probes, participants also rated their subjective level of situation awareness during each of the pauses in the driving simulation. Figure 5.5 presents the median self-reported SA score for each of the eight feedback conditions based on the 35 participants providing a total of 1164 data points. It can be seen that all of the median scores hover around the midpoint of the twenty-point scale, demonstrating neither strong nor weak self-perceived SA.

The Friedman test was performed and helps confirm the visual impression of the data shown above. No significant differences exist between subjective SA ratings according to feedback condition ($X_r^2 = 8.59$; df = 7; p = ns). In terms of effect size, $R_{phi} = 0.50$ which approximately translated

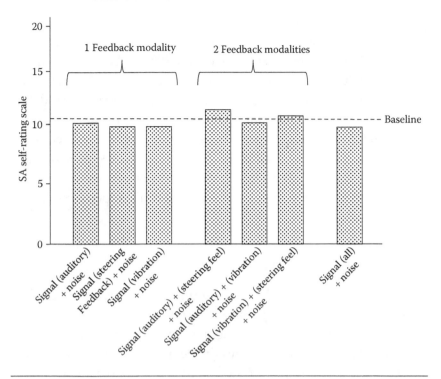

Figure 5.5 Median scores from self-report SA scale for each vehicle feedback condition ($N = 35$).

is indicative of a comparatively weak positive association between vehicle feedback as presented, and subjective SA as self-reported. The obtained R_{phi} value is directly comparable to the objective phase of the study (where $R_{phi} = 0.8$). It shows that although the overall effect is in the same general direction, self-perceived SA is not meaningfully related to the feedback presented through the simulator, at least at the 5% level. This finding occurs despite drivers being subjected to dramatic changes in feedback presentation, a difference of between one and four feedback modalities being present interchangeably. In other words, drivers felt more or less situationally aware across all conditions, despite very evident differences assessed via the probe recall technique, not to mention elsewhere (e.g. Chapter 2) via content analysis of on-road verbal transcripts.

5.3.2.1 Proving the Null Hypothesis The findings for subjective situational awareness are a paradoxical case of the null hypothesis (H_0) being regarded as the research hypothesis (H_a). Power analysis provides a means to address this conceptual hurdle (Cohen, 1990). The difference between the minimum and maximum values obtained is a difference along the twenty-point subjective SA scale of 1.8. Therefore, given the sample size and statistical tests were powerful enough to detect medium (or larger) effects with an 80% probability, and none were found, it would generally follow that differences of a magnitude greater than 1.8 would be required for a non-trivial effect to be identified (Cohen). It can be argued that changes along the scale of 1.8 or less are relatively trivial, especially in comparison to the effect sizes observed in the probe-recall phase of the study. It seems reasonable to conclude that overall driver self-awareness of SA, and more worryingly from a safety point of view, any shortfall in perceived SA, is very poor indeed.

5.4 Summary

Having seen in Chapter 2 that something has indeed happened to vehicle design, and there appear to be psychological and behavioural implications, this chapter has sought to extend the exploration in a controlled laboratory trial. The results demonstrate, firstly, that vehicle feedback is strongly associated with a driver's sensitivity to SA probe items. Secondly, the addition of non-visual vehicle feedback increases this sensitivity above vision alone. Thirdly, drivers are not self-aware of changes in their SA.

The findings support one of the prominent strategies suggested to enhance SA, which is to avoid overloading one sensory modality (such as vision) and instead spread the load across several (Endsley, 1988). These arguments also fit well with multiple resource models of the sort espoused by Wickens (1992) and others. In this study auditory feedback presents itself as the modality through which the most consistent increases in SA are seen. To some extent this echoes the work of Horswill and McKenna (1999) and Horswill and Coster (2002). To support good driver SA it appears the vehicle, as an intervening variable, needs to communicate the state of its environment, and its state within that environment, via auditory as well as visual means. Auditory feedback is a prominent aspect of noise, vibration and harshness (NVH) research, wherein the property 'noise' is generally regarded as unwanted. The effect of increasingly strict drive-by noise criteria, and a general desire for vehicles to exhibit the feel "of being in a larger, executive class car" (Bosworth et al., 1996, p. 115;

Becker et al., 1996) means considerable downward pressure has been placed on vehicle noise emissions. Interest now seems to be turning towards increasingly complex means to manipulate and shape noise in order to preserve 'desirable' aspects, whilst eliminating 'undesirable' ones (Barthorpe, 2000). SA provides a new dimension, offering a practical means to employ noise shaping technologies in the service of sustaining driver SA. Arbitrarily removing noise seems to place the attainment of optimum driver SA at risk.

Another key finding is that drivers' self-awareness of their own SA is poor. This is graphically shown by the drivers in the study being subject to very large changes in feedback presentation interchangeably, leading to a range of statistically significant differences in measured SA. Meanwhile, self-ratings of SA barely moved. Advanced management and quality approaches such as Quality Functional Deployment (QFD) are used by manufacturers to capture customer requirements (Akao, 1990), but QFD, and related approaches, tend to deal with psychological issues in a highly superficial and largely subjective manner. This study shows that the subjective information provided by consumers, and often used to drive processes such as QFD, does not necessarily capture hidden requirements like SA. In other words, consumers may not possess the level of self-awareness needed to drive important design decisions that potentially impact on their SA. The required level of awareness, therefore, falls to the vehicle designer.

To prevent such problems from going unseen, methods sensitive enough to detect SA decrements in a more objective manner under different design circumstances need to be used. SAGAT is a highly structured method, but does not meet the criteria for true objective measurement. SACRI is a valid enhancement to the SAGAT approach but is focused on a particular application domain. The revised method presented here charts a course between these two techniques; it builds upon the rigorous psychophysical structure offered by SACRI, and utilises the technique for eliciting SA requirements of a task from SAGAT. It attempts to avoid significant issues of multi-dimensionality in the SA probe items via a formal SA requirements analysis. Table 5.6 summarises the different approaches and shows how the best of both have been combined into something tractable and potentially useful for vehicle designers.

Table 5.6 Comparison of SA Measurement Techniques

	SAGAT	SACRI	REFINED
Use of measurable, publicly observable objective states	Some	Some	Yes
Probes derived from systematic HTA	Yes	No	Yes
Uni-dimensional probes	No	No	Yes
Ordinal rating scales	Yes	Yes	Yes
Use of non-parametric tests	No	No	Yes
Quantitative measurement structure	No	Yes	Yes
Generic method	Yes	No	Yes
Subjective method	Yes	Yes	Less so...

This study, in summary, paints an interesting picture. Returning to Lisanne Bainbridge's 'ironies' and Donald Norman's 'problems' it seems clear that as vehicle design becomes more advanced the contribution of the driver becomes more crucial, or at least more complex. Resolving issues of driver SA in the face of advanced vehicle designs will require even greater technological ingenuity. At the moment automotive engineering is at an intermediate level of intelligence and does not necessarily provide the continual feedback which occurs naturally in other circumstances. Inadequate feedback may well lead to difficulties for drivers but, even worse, it is a design opportunity that could be missed.

6

COMPARING CAR DRIVER AND MOTORCYCLIST SITUATION AWARENESS

6.1 Introduction

6.1.1 Human Factors and the Art of Motorcycle Maintenance

The previous insights and research questions were well received at the international conferences and meetings they were presented at (e.g. from Walker et al., 2001 to Stevens et al., 2016). It struck a chord with readers of *TEC* magazine (Walker, 2008), led to interesting newspaper headlines when it appeared in the popular press (e.g. 'Silent Killer Cars': *Manchester Evening News*, 15 Oct 2007), and stimulated a particularly intriguing conversation about the relative merits of classic W-Series Mercedes with a vehicle designer at Ferrari. One question arose repeatedly and it was this: "Wouldn't it be interesting to compare car driver and motorcyclist SA?" The questioners were frequently motorcyclists themselves (as are two of this book's authors) and all too aware of the dramatic differences in feel between the two vehicle types.

It was Robert M. Pirsig in the cult classic *Zen and the Art of Motorcycle Maintenance* who said:

> *"In a car you're always in a compartment, and because you're used to it you don't realize that through that car window everything you see is just more TV. You're a passive observer and it is all moving by you boringly in a frame. [...] On a motorbike the frame is gone. You're completely in contact with it all. You're in the scene, not just watching it anymore, and the sense of presence is overwhelming. That concrete whizzing by five inches below your foot is the real thing, the same stuff you walk on, it's right there, so blurred you can't focus on it, yet you can put your foot down and touch it anytime, and the whole thing, the whole experience, is never removed from immediate consciousness"* (1974, p. 14).

From a purely experimental point of view, setting aside Pirsig's inquiries into Buddhist philosophy, it was clear these two vehicle types provided an excellent naturalistic manipulation of feedback. Indeed, it was another opportunity not to be missed.

6.1.2 Cars versus Motorbikes

It hardly needs to be said that driving a car is different to riding a motorcycle. Motorcycles have hand controls for throttle, front brake and clutch (all of which are foot-operated in cars) and foot controls for gear selection (which are hand-operated in cars). Motorcycles are smaller, lighter, occupy less road space, are more manoeuvrable, faster and require balance, and in most cases afford an elevated view of the road compared to cars. Unlike car drivers, motorcyclists are directly exposed to the environment in which the machine is operating and to the noise, vibration and other sensations resulting from its interaction with the road.

Motorcycles also afford their riders with a direct and unassisted mechanical link between the primary control inceptors and the system under control. Instead of a steering wheel, steering column, flexible linkage, rank and pinion steering system, power assistance, track rods, ball joints and the relatively complex paraphernalia of a McPherson strut suspension layout, a motorcycle has handlebars and a front wheel affixed between two fork stanchions. Instead of a gear lever, or steering wheel mounted controls, an electronic double clutch and microprocessor-controlled automatic gear selection, a motorbike has a foot-operated gear lever protruding directly from its manual gearbox. Instead of a brake pedal, brake master cylinder, vacuum-powered servo assistance, microprocessor-controlled anti-lock brakes and complex handling management, a motorbike has a handlebar mounted lever with integral hydraulic piston applying pressure directly to hydraulic brake callipers. And so it goes on. All this, combined with significantly elevated power to weight ratios (approximately 115 bhp/tonne for a typical car versus 500 bhp/tonne or more for most motorcycles) confers a high degree of directness and responsiveness to the controls that is not replicated in all but the most extreme of 'supercars'. For many parameters we have an approximation to zero-order control, at least compared to the majority of cars. These factors combine to highlight the seemingly obvious point that driving a car is not the same as riding a motorcycle, and once again, it would be surprising if this did not have an effect on rider's SA.

No previous studies have been identified which contend with this issue. Previous studies that tackle motorcycling tend to focus on rider skill (e.g. Rice, 1978), machine dynamics (e.g. McKibben, 1978; Hartman, 1978), safety (e.g. Hurt and DuPont, 1977; Patel and Mohan, 1993) and anthropometrics (Robertson and Minter, 1996). Motorcycles, therefore, represent a unique experimental opportunity. Motorcycles appear to offer an independent variable of sufficient strength to be able to observe the effects of vehicle design and feedback in relation to driver SA in naturalistic settings. Given that such a study has not previously been attempted (at the time of writing at least) the findings promise to be fascinating.

6.1.3 Distributed Situation Awareness

Alongside the introduction of a different vehicle type is an expansion of the situation awareness (SA) concept itself. Rather than persist with the individualistic view of SA deployed in the previous chapter, the opportunity is now taken to widen the system boundaries and insert the team and systems lens in front of the debate. As we have seen already, researchers have made powerful arguments for more systems thinking in Human Factors (Walker, 2016; Walker et al., 2017; Dul et al., 2012), perhaps no more so than Hutchins' (1995a) pioneering work on distributed cognition. In some ways this goes furthest by placing inanimate, non-human, technical entities on the same footing as humans within a joint cognitive system. This should not be an entirely shocking idea bearing in mind many of the fundamental concepts in cognitive psychology stem from cybernetics and the burgeoning computer-science field. In other words, some of the original cognitive concepts were often applied first to automated control systems rather than humans. In practice the idea of collaborative human and non-human agents extends SA beyond the individualistic perspective into the realm of distributed situation awareness (DSA) (Stanton et al. 2004, 2017). Road transport, moreover, provides an excellent demonstration of the concept.

On the one hand, situationally aware drivers become 'aware' based on information received from the environment, with the vehicle often serving as an intermediate variable. Thinking more broadly it is also the case that situationally aware drivers, and riders, also manage to communicate their intentions to other road users (and anticipate the intentions of others) without verbal communication. This is partly due to the constraints of the system and the conventions of road use, and partly due to compatible, but non-identical forms of SA. No one driver in the traffic stream will have the same SA as any other driver; each will be unique. Despite, or rather because of this, the compatible and behaviourally interacting SA enables the loosely coupled road transportation system to function effectively. What is particularly clear in the comparison between car drivers and motorcyclists is that degradation in the SA of any one 'agent' in the system may be compensated for by other agents. Thus the system is both dynamic and flexible. This gets to the heart of how two very different vehicle types (cars and motorcycles) can co-exist in the same environment despite widely differing SA requirements.

An important concept within the DSA approach is that the vehicle and other aspects of the surrounding context will possess their own SA capability. The vehicle deals with stresses it experiences as it traverses the environment, transforms those stresses through its mechanical systems, and 'displays' them to the driver via vehicle feedback. The traffic stream itself displays its status to the drivers within it via the proximity and configuration of other vehicles, a configuration that affords some types of behaviour while constraining others. Road signs represent information in a very direct way, while road geometry may do so more indirectly. Discrete instances of information within this system of road users, road infrastructure, wider environment etc. can be referred to generically as information objects. The term refers to an "entity or phenomenon about which an individual requires information in order to act effectively" (Baber, 2004). From an individual perspective, SA could be defined as "activated knowledge [that] a person has about a dynamic scene [or scenario]" (Gugerty, 1998, p. 498). This is what the normative probe recall methods used in the previous chapters are accessing. From a systems perspective, however, SA can be viewed more broadly and formatively as "the sum of knowledge relating to specific topics within the system" (Baber, 2004). A different approach is needed to access these.

There are serious methodological constraints placed on the process of SA measurement by the naturalistic experimental context. A freeze-probe methodology of the sort describe in Chapter 5 is not practical to implement in this setting. The dynamic road environment, including other traffic, cannot be easily 'frozen' then 'restarted'. Quite aside from which is the more philosophical point that in freezing a situation one is interrupting the object of real interest, the flow of situations and cognition. According to psychologist William James, one would not study the flow of a river by extracting buckets of water from it. The alternative in this case is to deploy concurrent verbal protocols (Ericsson and Simon, 1993) in a manner similar to the first study presented in Chapter 2. In this case drivers and riders provide a formative 'running commentary' as they control their vehicle. The transcripts of the driver's/rider's verbalisations are then encoded and categorised in order to extract the relevant information objects. These can then be connected to form a network of information representing the system's SA, including the driver's individual SA.

The exploratory hypotheses to be explored in this chapter are two-fold: firstly, what can the study of motorcyclist SA tell us about the extent to which drivers might be isolated from their environment? Secondly, how does this affect the behaviourally interacting nature of these two road users? A comparison between cars and motorbikes also permits the DSA approach to be introduced and for driver and motorcyclist SA to be explored in a unique way.

6.2 What Was Done

6.2.1 Design

The experiment was exploratory and based upon a naturalistic on-road driving paradigm in which a sample of car drivers and motorcyclists used their own vehicles around a defined course on public roads. The experimenter travelled in the front passenger seat during the observed runs in the cars, or followed on another motorcycle during the observed runs with the motorcycles.

Previous research suggests that observation produces reliable and valid data and that driving style is not affected to a significant degree (Hjalmdahl and Varhelyi, 2004; Salmon et al., in press). The use of the driver's/rider's own vehicle also ensures a degree of familiarity which is difficult to achieve with an experimental vehicle. It is a step taken to ensure any sensitivity drivers exhibit towards vehicle feedback is given the best opportunity to emerge experimentally.

Drivers/riders were instructed to verbalise the information, objects and artefacts within the driving context they were attending to. They were not required to attend only to specific information types or attempt to provide a commentary on processes normally undertaken in an automatic fashion. The benefits of this approach are that it minimises the intrusiveness of the verbal protocol on the primary task of driving/riding and helps ensure the content and sequence of information processing is relatively unaffected by performing the commentary task.

Information objects relating to the driver's/rider's moment-to-moment SA were extracted from the transcripts of the commentaries using content analysis. Inter- and intra-rater reliability were checked against two independent analysts. The verbal protocol was dependent upon one between-subjects independent variable (vehicle type) with two levels: car or motorcycle. The between-subjects paradigm required several controlling measures to ensure vehicle type was the most systematic experimental manipulation in amongst a fairly noisy naturalistic data environment. The controlling measures were self-report questionnaires of driving style, recordings of average speed and time, and demographic data. In addition, all experimental trials took place at predetermined times to control for traffic density and weather conditions.

6.2.2 Participants

Twelve participants took part in the study using their own vehicles. They were comprised of six car drivers and six motorcyclists. All participants held a valid UK driving licence with no major endorsements. All participants reported they drove approximately average mileages per year for their vehicle type. The participants fell within the age range of 20 to 35 years old. Mean driving experience was 5.83 years for the car drivers and 5.33 years for the motorcyclists. Both groups can be considered closely matched on these dimensions and have been exposed to many hundreds of hours of driving or riding experience, with at least one year of familiarity with their current vehicle. Once again, the challenges of recruiting female participants arose – something that has been a source of frustration in our studies for some time, and a worrying indication of the transportation field more generally. Despite this, within the context of a small sample size gender differences could provide an unwanted confounding variable. In addition, motorcycles comprise only 3% of registered vehicles on British roads, the majority of which are ridden by males (Chorlton and Jamson, 2003). As a result, an all-male sample was used.

6.2.3 Materials

Instructions and Questionnaires: Driving style was assessed via the Driving Style Questionnaire (DSQ) (West et al., 1992). In this case question items referring to the 'deviance' scale were removed due to an incompatibility with motorcycle riding. Certain question items referred to behaviours which were legitimate for motorcyclists but not car drivers, for example, filtering through traffic, certain forms of overtaking etc. Standardised instructions concerning the desired form and content of the verbal protocol were provided for each participant. Written instructions were also devised for the protocol analysis encoding scheme. This was for use by the experimenter during the analysis phase, and by additional coders used to test inter- and intra-rater reliability.

Vehicles and Apparatus: Six cars (a Volkswagen Golf TDi, Audi TT, Toyota Tercel, BMW 325i, Volkswagen Golf CL and Peugeot 309 GLD) and six motorcycles (Triumph Daytona 900, Suzuki TL1000R, BMW R1100GS, Laverda 750 Formula S, Suzuki GSX400F and Honda CBX750) were used, all of UK specification. Some of the cars were drawn from the 'over-sampled' participant pool developed in Chapter 2. Car drivers were videoed whilst they drove using a miniature video camera and laptop computer. Motorcyclists were audio-recorded using a microphone mounted in the chin-piece of their crash helmet and a digital recording device carried on their person. An identical setup was used for the accompanying rider who verbally annotated the ride by reporting on major contextual events and start/finish points. The two recordings were merged to form the final experimental transcript.

The Test Route: The on-road route was the same as that used for the study described in Chapter 2. It is contained within the West London area of Surrey and Berkshire and is 14 miles in length not including an initial three-mile stretch used to warm up participants. The route involved an approximate driving/concurrent verbal protocol time of 30 minutes, thus helping to reduce fatigue effects. The route enabled all national speed limits to be attained and was comprised of one motorway section, seven stretches of A or B classification roads (trunk roads), two stretches of unclassified roads (minor roads), three stretches of urban roads, one residential section and fifteen junctions. Experimental runs took place at 10:30 in the morning, 2:30 in the afternoon (Monday–Thursday), 10:30 on Friday, or any daylight time during weekends. These times avoided peak traffic hours for the area and were all completed in dry weather.

6.2.4 Procedure

Formal ethical consent was obtained from all participants before the study commenced. Particular emphasis was given to the participant's responsibility to ensure their safety and that of other road users. Participants then completed the DSQ followed by a comprehensive experimental briefing. The concurrent verbal protocol consisted of the driver providing a 'running commentary' about the information they were taking from the driving scenario and how they were putting it to use. An instruction sheet on how to perform a concurrent verbal protocol was read by the participant, and the experimenter provided verbal examples of the desired form and content. In the case of the motorcyclists, they were further instructed that the experimenter would follow them on another motorcycle in an offset road position. They were instructed to use their mirrors as normal, watching for directional indications from the experimenter, and to act upon them. The three-mile approach to the start of the test route enabled participants to practise and to be advised on how to perform a suitable concurrent verbal protocol. During the data-collection phase the in-car experimenter remained silent apart from offering route guidance and monitoring the audio/video capture process. For the motorcyclists the experimenter followed at a safe distance, remaining in the lead rider's rear-view mirrors by riding in an offset position, and using their own indicators to guide the participant around the route. The experiment was complete upon return to the start point.

6.3 What Was Found

6.3.1 Analysis of Control Measures

Table 6.1 presents the descriptive, inferential and effect size analysis of the median speed around the experimental road course, and the outcome of the DSQ.

No statistically significant difference was detected for DSQ scores between motorcyclists and car drivers (U(N1 = 5, N2 = 5) 9, p = ns). Statistical power is not sufficiently high to presume there is literally zero difference. The effect size correlation R_{bis} suggests that approximately 2% of the variance in DSQ scores is explained by vehicle type, and therefore a function of the type of people who choose to use these vehicles. However, if this finding is a genuine population effect then it is small enough to be of little practical importance for current purposes. On the other hand approximately 21% of the variance in median speed is explained by vehicle type and the difference between the two groups is statistically significant, at least at the 9% level (U(N1 = 6, N2 = 6) 7.5, p < 0.09). Motorcyclists are completing the course more quickly and spending slightly less time on it. Experience out on the test route indicates that most of this time was gained by being able to overtake more freely and to filter through the occasional traffic queue. Only in one case were motorcycle speeds appreciably higher than the car speeds. Despite all this, the variance overall is still under 5 mph. Combined with the additional weather and traffic controls put in place, the two groups can be regarded as tolerably well matched,

Table 6.1 Summary of Analysis for the Control Measures

	MEAN BIKE	MEAN CAR
Lap Speed	34.83 MPH	30.99 MPH
Mean Overall DSQ Score	3.93	3.46

	TEST STATISTIC (U)	PROBABILITY	EFFECT SIZE (R_{bis})	VARIANCE EXPLAINED (R^2)
Lap Speed	7.50	$P < 0.09$	0.46	0.21 (21%)
Mean DSQ Score	9.00	$P = 0.47$	0.13	0.02 (2%)

certainly to a sufficient degree to enable driver/rider SA differences to emerge should they be strong enough to do so.

6.3.2 Analysis of Concurrent Verbal Protocol Data

6.3.2.1 Data Reduction The verbalisations captured by the audio/video equipment were transcribed verbatim against a two-second incremental timeline. A theme-based analysis was then undertaken centring on the extraction of content words from function words, as detailed in Chapter 2. The encoding scheme was also grounded according to what part of the 'driving system' they referred to. Under this scheme the encoding is 'exhaustive' and non-mutually inclusive. Whenever a content word meets the definitions described in the encoding instructions it enters into the analysis, irrespective of whether it is already represented in other categories.

After completing the encoding process for all participants, the reliability of the encoding scheme was then established. Two independent raters encoded previously encoded analyses in a blind condition using the same categorisation instructions the original rater employed. Across the seven individual encoding categories inter-rater reliability (IRR) was established at $R_{ho} = 0.7$ for IRR 1 ($n = 756$) and $R_{ho} = 0.9$ for IRR 2 ($n = 968$), both these values being significantly correlated ($p > 0.05$ for IRR 1 and $p > 0.01$ for IRR 2). Intra-rater reliability was also examined to check for any drifting in encoding performance over time. This analysis revealed a correlation of $R_{ho} = 0.95$; $n = 756$; $p < 0.01$, suggesting that encoding performance over time is remaining stable. At this exploratory level of analysis the reliability of the encoding scheme seems reasonable.

6.3.2.2 Rate of Verbalisations A total of 8065 information objects were identified from within the transcripts provided by car drivers and motorcyclists. Motorcyclists provided a total of 5022 information objects compared to 3043 for car drivers. This difference is significant, as revealed using a binomial test ($p < 0.01$). If SA is viewed simply as the sum of knowledge relating to specific topics within the system (e.g. Baber, 2004), then this represents evidence motorcyclists may be extracting more information from the scenario and/or using the same information more often. This finding is further strengthened by the fact motorcyclists spent significantly less time around the course, providing this increased reporting of information objects in less time than car drivers.

6.3.3 Results of Systemic SA

6.3.3.1 SA Content Verbalisations were categorised according to what parts of the system different information objects came from or referred to. The categories were similar to those described in Chapter 2. They were: Own Behaviour, Behaviour of the Vehicle, Road Environment and Other Traffic. Under the precepts of DSA, these categories would specify four 'specific topics' or system elements, and the category of information objects that drivers/riders are extracting from them. Examples from verbal transcripts include "INDICATE left, CHECK all MIRRORS" [behaviour], "CAR just sitting at 2000 REVS [vehicle], "SIGNS say ONE WAY to the left" [road], and "TRAFFIC going SLOW" [traffic]. The capitalised words show the information objects. Figure 6.1 below presents the results as a DSA profile. It shows the content of SA is not identical for the users of the two vehicle types.

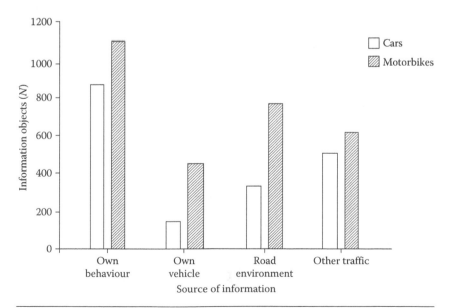

Figure 6.1   Systemic SA profile of cars and motorcycles ($N = 12$).

Statistically significant differences across the profile of system elements were detected for motorcycles and cars using a Friedman test (Chi Square = 14, df = 3, p < 0.01 and Chi Square = 13, df = 3, p < 0.01 respectively). Multiple comparisons (using a procedure outlined in Siegel and Castellan, 1988) reveal a pattern of significant differences that differs according to vehicle type (Table 6.2).

The main differences are that car drivers extract significantly more information objects from the vehicle and the road than they do from other traffic (at the 1% and 10% level respectively). Motorcyclists extract significantly more information objects from the road environment than they do from their vehicle, and talk about their own behaviour significantly more than that of other traffic (whereas car drivers do not). These results illustrate some key differences in the content of SA possessed by motorcyclists and car drivers. This can be explored in more detail using an approach that focuses on key information objects within the DSA network.

The first step is to combine the verbal transcripts for all participants in the study and subject this master list to a word frequency analysis. From this it is possible to derive an ordered hierarchy of key information objects from a total list that numbers well over 200. The top 16 are presented in Table 6.3. After the sixteenth rank, the number of times any information object is mentioned by drivers/riders diminishes rapidly. Thus it represents an 'elbow' in the graph, and a sensible point at which to differentiate between key information (mentioned often and consistently by all participants) and subsidiary information (mentioned infrequently and sporadically). The data is then split into driver and motorcycle cohorts and the same analysis repeated so that the top 16 information objects relevant to each vehicle type can be compared against each other and the master list.

Table 6.2 Summary of Multiple Comparisons across the Profile of Systemic SA.

COMPARISONS	MEAN RANK BIKES	MEAN RANK CARS
Behaviour-Vehicle	2.67 (p < 0.01)	2.34 (p < 0.01)
Behaviour-Road	1.33 (p < 0.1)	1.84 (p < 0.01)
Behaviour-Traffic	2.00 (p < 0.01)	0.50 (p = ns)
Vehicle-Road	1.34 (p < 0.1)	0.50 (p = ns)
Vehicle-Traffic	0.67 (p = ns)	1.84 (p < 0.01)
Road-Traffic	0.67 (p = ns)	1.34 (p < 0.1)

Table 6.3 Analysis of Key Information Objects

KEY INFORMATION OBJECTS	MOTORCYCLES	CARS
Road	■	■
Right	■	■
Left	■	■
Traffic	■	■
Car	■	■
Lane	■	■
Ahead	■	■
Speed	■	■
Gear	■	
Braking		■
Roundabout	■	■
Corner		■
Revs	■	
Indicating		■
Lights	■	■
Junction	■	■

It is clear the top eight information objects (which all seem to relate to a form of spatial SA) are shared between the two vehicle types. Moving down the hierarchy the pattern starts to diverge. Even when considering only the top 16 objects, it can be noted 31% of them differ between vehicle type, and continue to diverge as the depth of analysis (and number of information objects) is compared. What we have here, then, is an example of shared 'identical' SA across *some* information objects, consistent with traditional notions of 'team SA' (e.g. Endsley and Robertson, 2000). But, in addition, there are clearly shared as in 'compatible but different' SA objects, consistent with other views of team SA (e.g. Shu and Furuta, 2005).

6.3.3.2 SA Quantity Statistically significant differences were detected between cars and motorbikes in the behaviour category (U(N1 = 6, N2 = 6) 8, $p < 0.1$), the vehicle category and the road category (U(N1 = 6, N2 = 6) 3, $p < 0.05$). Motorcyclists are reporting significantly more knowledge objects than car drivers in all of these system domains (Figure 6.2). This is accompanied by large effect sizes. The most interesting finding is the vehicle category. This clearly illustrates the extent of difference between the two vehicle types in terms

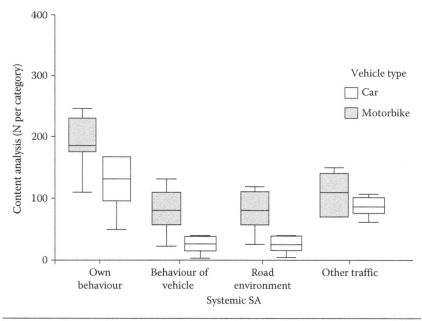

Figure 6.2 Quantity of SA across the systemic categories for cars and motorbikes.

of feedback and its effect on driver SA. The hypothesis concerning feedback and isolation appears to be borne out by these results. The comparative isolation car drivers might experience relative to motorcyclists is certainly represented by a reduced set of information objects (Table 6.4).

6.3.3.3 Conversion of Information into Knowledge A consistent finding above is that drivers talk a great deal about the state of their own SA and the related information objects, decisions and actions that flow from it. The interest now, therefore, turns towards the means by which individuals acquire the state of SA, linking the systems approach back to an individual one. To do this involves taking the extracted information objects and categorising them into Level 1 (the simple perception of elements in the environment), Level 2 (comprehension as to what those elements mean for the current task) and Level 3 SA (projection of this understanding onto planned and future actions) (Endsley, 1995). This process of re-categorisation was also checked for inter-rater reliability. Figure 6.3 presents the outcome of this analysis for car drivers and motorcyclists.

Table 6.4 Summary of Analysis on the Quantity of SA According to Vehicle Type

	BEHAVIOUR	VEHICLE	ROAD	TRAFFIC
Non-Parametric Statistic (U)	8	3	3	13.5
P-Value	0.11	0.02	0.02	0.47
Approximate Effect Size (R_{bis})	0.44	0.55	0.55	0.12
Variance Explained (R^2)	0.19 (19%)	0.30 (30%)	0.30 (30%)	0.01 (1%)

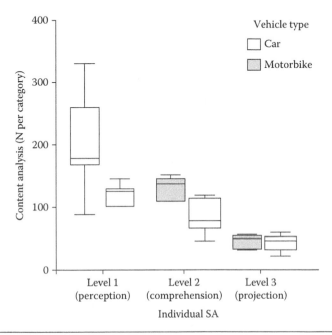

Figure 6.3 Summary of individual driver SA according to vehicle type and the three-stage model of SA.

Figure 6.3 presents some key differences in the processes by which the drivers and riders acquired the state of SA. Significant differences were detected in Level 1 SA (U(N1 = 6, N2 = 6) 6, $p < 0.6$) and Level 2 SA (U(N1 = 6, N2 = 6) 4, $p < 0.05$). Both Level 1 and 2 SA are more dominant for motorcyclists. This finding is perhaps logical given the extra and/or more frequent use of the same information perceived and comprehended from the vehicle and road environment. Once again, the effect size correlation is suggestive of a large effect (Table 6.5). It is not fully clear whether car drivers are significantly disadvantaged in Level 1 and 2 SA terms compared to motorcyclists, or if car drivers are able to project future states on less frequent occurrences of

Table 6.5 Summary of Analysis on Individual Driver SA According to Vehicle Type

	PERCEPTION	COMPREHENSION	PROJECTION
Non-Parametric Statistic (U)	6	4	15
P-Value	0.06	0.03	0.63
Approximate Effect Size (R_{bis})	0.49	0.53	0.00
Variance Explained (R^2)	0.24 (24%)	0.28 (28%)	0.00 (0%)

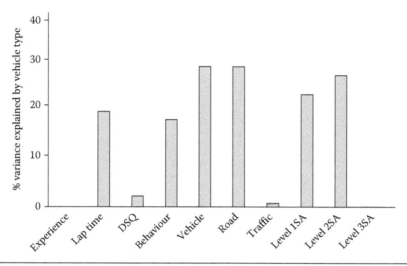

Figure 6.4 Effect size summary showing the % of variance explained in each of the control and dependent variables according to vehicle type.

Level 1 and 2 SA. In either case, the suggestion there might be some kind of adaptive process in play does not seem far-fetched.

6.3.4 Effect Size Summary

Figure 6.4 presents an effect size summary showing the control and dependant variables, and the approximate amount of variance explained by vehicle type. It suggests that vehicle type explains close to 0% of the variance in the control variables of experience and driving style. Much larger effect sizes are in evidence for the independent variables of systemic and individual SA. Motorcycle rider SA is clearly different to car driver SA.

6.4 Summary

This was a fascinating study which provided information on the nature and quantity of DSA; demonstrated it could be measured;

and showed how two theoretical lenses could be applied to the same dataset to expand the range of insights. Returning to the question that was so often asked at the conclusion to conference presentations, what do motorcycles tell us about isolation from vehicle feedback? In short, the hypothesis that motorcycles provide greater feedback than cars seems justified. The SA of motorcycle riders is significantly 'greater'. But does this mean 'more' is 'better' and that cars should be trying to emulate motorcycles in some way? Perhaps not. What the results do communicate is when the direct linkage between primary vehicle controls is mediated by an increasing array of technologies and systems driver SA is significantly affected. Motorcycles afford their rider with a very direct linkage between the control inceptors and the resulting vehicle outputs compared to cars, and the results can be seen in quantitatively and qualitatively different states of SA. This, in turn, could have far-reaching effects on behaviour.

While a return to motorbike levels of feedback in cars might not be advisable, it is certainly the case that an increasing divergence is taking place. The trend in vehicle comfort and amenity is likely to isolate drivers even more than is currently the case. The outcomes of this study seem to hint at the idea drivers can adapt, generating comparable Level 2 and 3 SA on reduced Level 1 SA. What is currently unknown is how well drivers will be able to continue adapting, what the adaptive limits are, and what effect this will have on the loosely coupled road transport system as a whole. This is a theme developed further in the next chapter.

7

COGNITIVE COMPATIBILITY

7.1 Introduction

7.1.1 Gulfs of Evaluation

It makes intuitive sense that motorcyclists will interpret the same road situation differently to car drivers. The question this chapter explores is whether this can really be regarded as the case, and if so, whether car drivers and motorcyclists can be regarded as cognitively compatible. If they are compatible then safety interventions which foreground the objective state of the situation (i.e. increased rider conspicuity) will be more likely to have an effect. This is because drivers will be interpreting the situation in a way that is already favourable to the anticipation of other road users, and additional conspicuity (for example) fits the schema. If, however, they are incompatible, then a more nuanced approach might be needed. For example, no matter how visible a rider may be, if the driver is operating in a situation which is generating a strong stereotypical response which is unfavourable to the observation of other road users, then in order to improve safety changing the mental representation of the situation becomes as important as its objective state.

A cognitive incompatibility might be described by Norman (1990a) as a 'gulf of evaluation'. This describes a person's attempt to make sense of their context and how it matches their expectations and intentions. Difficulties arise because, quite often, designers and users of a system bring to bear different cognitive models of a system based on their own understanding of it. This leads to incompatibilities between what the designer expects and what the user wants. It is not too much of a stretch to say that similar 'gulfs of evaluation' could exist in terms of how identical road situations are interpreted, and what those situations might 'afford' for different road users.

The concept of 'affordances' reflects Gibson's (1977, 1979) idea of a relationship existing between people and their immediate context.

It also reflects Neisser's (1976) concept of the environment being sampled, which in turn modifies behaviour, which in turn guides further exploration. Once more, it would be surprising if users of vehicles with very distinct differences in feedback – such as cars and motorbikes – did not experience the same road environment differently. This is certainly the case for visual search patterns (e.g. Underwood et al., 2003; Hosking et al., 2010) and in this chapter we explore whether it is also the case for situation awareness (SA).

Incompatibilities between different groups of road users are cited as one of five key road safety problems persistent over time, between nations, and not easily solved (Elvik, 2010). Problems at the motorcycle and other-road-user interface are well rehearsed in the literature. A particular problem is how other road users perceive motorcyclists (Clarke et al., 2007). This problem is rendered all the more interesting because drivers who are more likely to have difficulty interpreting a road situation involving an oncoming motorcyclist tend to be older, more experienced drivers who would ordinarily be expected to interpret a situation correctly. Another interesting finding is that while car drivers and motorcyclists seem to have the most problems interacting with each other at junctions, motorcyclists seem to encounter more problems with the road itself, particularly geometric features like curves and bends (Clarke et al., 2007; Daniels et al., 2010). Perhaps the clearest evidence of cognitive compatibility issues between car drivers and motorcyclists is provided by Magazzù et al. (2006) who found that car drivers who also held motorcycle licenses were less likely to be responsible for motorcycle–car crashes than drivers with car-only licenses.

What previous studies seem to show is that identical roads can be experienced differently by different road users, and this interface can be mutually reinforcing or conflicting. From an experimental perspective, motorcyclists offer an independent variable of sufficient strength to be able to observe such effects directly in a naturalistic setting. Given increasing ridership (aided by congestion charging, environmental concerns and other societal factors which favour motorcycle usage) and the disproportionate number of motorcycle accidents (e.g. Clarke et al., 2004), such opportunities enable a much needed insight into novel accident prevention strategies.

7.1.2 Less Could Be More

Situation awareness and the notion of cognitive compatibility both infer a "mapping of the relevant information in the situation onto a mental representation of that information" (Rousseau et al., 2004, p. 5). The term 'mental representation' reflects two further aspects of situation awareness. Firstly, a mental 'representation' infers that information elements are structured in some way. A mental representation is not merely about the presence or absence of discrete elements of awareness but also about their interconnection. This leads to a second and more fundamental point. A mental representation also reflects the "hypothetical nature of perceptual experience" (Bryant et al., 2004, p. 110). In other words a mental representation, or the 'awareness' of a situation, is a model (e.g. Banbury et al., 2004), "a representation that mirrors, duplicates, imitates or in some way illustrates a pattern of relationships observed in data or in nature [...]", or "a characterisation of a process [...]".

A mental representation needs, foremost, to provide the rider or driver with "explanations for all attendant facts" (Reber, 1995, p. 465, 793). This, in turn, does not necessarily require a particularly rich or detailed model of the situation; indeed, it would be a highly inefficient representation if it did. On the contrary, there is good evidence that the better the mental representation, the more parsimonious it is (e.g. Gobet, 1998; Chase and Simon, 1973). It needs to be. Estimates vary, but the neuroscience literature suggests the human brain can store between 1 terabyte and 2.5 petabytes of information. Given the inflow of information from the driving context this upper limit would be reached quite quickly if the brain did not find a way to cleanse the data it is provided with, and compress it for storage (to use a computer science metaphor). It would also take longer to access if it were stored as if on video tape. This, again, would be a problem for fast-moving dynamic tasks like driving. Clearly the more parsimonious the mental representation the more it is 'cleansed' and 'compressed' and the more that raw information elements are combined into higher levels of abstraction or themes.

The relationship between the quantity of information contained in a given mental representation and its interconnection gives rise to a number of interesting characteristics. An example is described by Metcalf's Law (e.g. Metcalf, 1973). This states that the configuration of any given quantity of information embodies a property called 'value'. Information elements increase in value the more they are linked to other information elements. According to Metcalf's Law, as the number of links between individual elements increases linearly, the value of the entire configuration of elements and interconnections, or the rider's/driver's mental representation, increases exponentially. In other words, both the structure and content of SA are important.

7.1.3 SA as a Network

The representation of knowledge in a network is not a new concept. Semantic networks have been used by cognitive psychologists since the 1970s. Semantic networks are based on the long-held belief that all knowledge is in the form of associations. More sophisticated forms of semantic network are based explicitly on Ausubel's theory of learning (Ausubel, 1963). This suggests meaningful learning occurs not by the simple addition of new concepts to existing concepts, but by their assimilation. This can be represented by depicting linked nodes in a network (Eysenck and Keane, 1990). Assimilation would occur not just by the addition of new nodes in a network, but by increasing the strength of pre-existing connections and the formation of new ones. The previous chapter began to deal with the notion of information networks and we can develop these ideas further here.

Within a semantic network (e.g. Figure 7.1) each node represents an object. Nodes are then linked with edges (lines) typically specified by verbs or by other more sophisticated means. By way of a simple example we might say this object 'has' the property of this other object, 'has' denoting a link should exist between objects/nodes. Anderson (1983) built on the idea of semantic networks, and their

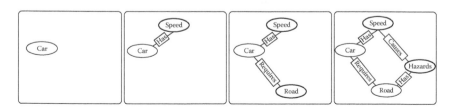

Figure 7.1   Example of a semantic network.

associated learning theories, to describe activation in memory using 'propositional networks'. Salmon et al. (2009) and Stanton et al. (2009) have since extended this approach into the realm of situation awareness and have anchored it successfully to a generative model of cognition (e.g. Neisser's perceptual cycle, 1976) and to Schema Theory (Bartlett, 1932). These help to describe how individuals possess mental templates of past experiences (i.e. schemas) which are mapped with information in the world to produce appropriate behaviour (i.e. a generative model of cognition). As described earlier, a schema is rather like a mental template. It is neither completely new behaviour nor merely a repetition of old behaviour, but is behaviour generated from a familiar set of initial conditions, both mental and physical. Schema Theory offers an explanation for the paradoxical case described above in which more experienced drivers seem to have greater degrees of cognitive incompatibility with motorcyclists. In this case, because cars are more numerous than motorcycles (in the UK at least), repeated experience with the latter generates strong stereotypical behaviours potentially unfavourable to the former. In other words, the links between concepts relating to 'cars' are likely to be stronger (because they are experienced more often) than those relating to motorbikes.

The concepts of Genotype and Phenotype Schemata are particularly relevant here. Genotype Schemata can be thought of as the 'global prototypical routines' contained in the mind of the person. They help define a person's propensity to behave in a certain way in a situation containing certain features. Phenotype Schemata refer to local 'state specific' routines, or activated schema brought to bear on a specific problem in a specific situation. Hollnagel (1993), for example, uses a similar distinction to illustrate how generic error modes (the genotype) may be related to specific observed errors (the phenotype) in the world. Based on this view it could be stated that the root of cognitive incompatibility between different road users are faults in triggering appropriate schema. Norman (1981) puts forward three such faults which account for the majority of errors. These are:

1. Activation of wrong schemata due to similar trigger conditions, such as 'car not present therefore I can pull out', despite there being a motorcyclist actually present.

2. Failure to activate appropriate schemata due to a failure in detecting trigger conditions indicating a change in the situation, such as a motorcycle approaching rather than a car.
3. Faulty triggering of active schemata. In other words, triggering the appropriate schemata either too early or too late to be useful, such as stopping the car beyond the stop line and causing the motorcyclist to change their speed or trajectory.

The use of networks, and their associated theoretical concepts, are well established in the field of psychology and the human sciences more generally. Good progress has been made already using human factors methodologies to populate such networks and draw inferences from them. A novel development described in this chapter is the use of a sophisticated software tool called Leximancer™ which automates the creation of such networks, and does so with complete repeatability. This represents the next step in adopting a more systemic view of SA. Leximancer™ uses text representations of natural language, such as transcripts from concurrent verbal protocols produced by riders and drivers, to create networks of interlinked themes, concepts and links. This is achieved by sophisticated algorithms based on word proximity, quantity and salience. Leximancer™ has been used extensively in previous studies. It has provided insight into organisational change (Rooney et al., 2010), intergroup communication (Hewett et al., 2009), analysis of web content (Coombs, 2010) and large scale meta-analyses of themes within scientific journals (Cretchley et al., 2010). The application of this technique to a more intensive form of analysis based on concurrent verbal protocols of real-life transport contexts is a novel one.

7.1.4 Exploratory Analysis

The discussion above establishes a relationship between semantic networks, generative models of cognition, Schema Theory and, with an appropriate systems lens inserted, SA. It means that properly constructed semantic networks can serve as a useful analogue for aspects of a road user's mental representation of a situation. It is from this theoretical basis that we can proceed into the generation of more practical insights. The first of these is to analyse the question of whether vehicle type (car versus motorcycle) does actually affect an individual's experience of the same, or very similar, road situations. To do this, semantic networks were created based on concurrent verbal transcripts performed at the time the situation was experienced by riders or drivers. Cognitive compatibility can be assessed by analysing the structure and content of those networks. According to distributed situation awareness (DSA) (Salmon et al., 2009; Stanton et al., 2009) it is perfectly feasible for there to exist pronounced differences in network content and structure, but for those differences to be compatible and mutually reinforcing. It is equally feasible for differences in content and structure to be incompatible, with the behavioural effects of one mental representation not compensated for by the other. The study presented in this chapter proceeds in an exploratory manner, trading strict experimental control for ecological validity. The study is intended to build on the systems SA agenda initiated in the previous chapter.

7.2 What Was Done

7.2.1 Design

The experiment employs the same participants, high-level design, materials and procedure as described in Chapter 6. It is an exploratory study based on a naturalistic on-road driving paradigm in which individuals use their own vehicles around a defined course on public roads. The experimenter travelled in the front passenger seat during the observed runs in the cars, or followed on another motorcycle during the observed runs with the motorcyclists. This controlled for the possible effects of observation upon driving behaviour, as explained previously. Drivers/riders were required to provide a concurrent verbal protocol as they traversed the road course, which was then analysed using a text analysis tool called Leximancer^{TM} (see Smith, 2003). This enabled differences in textual and thematic content to be systematically analysed, and the structure of the verbal protocol to be represented via semantic networks. These outputs were dependent on two independent variables: vehicle type and road type. Vehicle type had two levels: car or motorcycle. Road type had six levels: motorway (freeway), major road, country road, urban road, junction and residential road. Controlling measures were self-report questionnaires of driving style, recordings of average speed and time, and demographic data. All experimental trials took place at defined times to control for traffic density and weather conditions, as noted earlier in Chapter 6.

7.2.2 Procedure

The procedure was identical to that described in Chapter 6 apart from the following:

The verbal protocol data collected from the drivers and riders was, in this case, treated with Leximancer™. This is a software product which automates the process of semantic network creation. Six main stages are performed in order to transform the verbal transcripts into semantic networks:

1. Conversion of raw text data (e.g. definition of sentence and paragraph boundaries etc.).
2. Automatic concept identification (e.g. keyword extraction based on proximity, frequency and other grammatical parameters).
3. Thesaurus learning (e.g. the extent to which collections of concepts 'travel together' through the text is quantified and clusters formed).
4. Concept location (e.g. blocks of text are tagged with the names of concepts which they may contain).
5. Mapping (e.g. a visual representation of the semantic network is produced showing how concepts link to each other).
6. Network analysis (this stage is not a part of the Leximancer™ package but was carried out as an additional step to define the structural properties of the semantic networks).

Whereas the study described in Chapter 6 focused on defining and exploring the information nodes, the use of Leximancer™ in this chapter enables their interconnection to be explored. This, then, extends the systems approach to SA further.

7.3 What Was Found

7.3.1 Extracting Semantic Content from the Concurrent Verbal Protocols

A metric for the amount of semantic content able to be extracted from different road scenarios is given by the word count of the verbal transcripts. The total word count across all road types and both road users is 28,169. Under the null hypothesis the total word counts for motorcyclists and car drivers should be 14,084 (i.e. 28,169 / 2). In fact the findings show the total word count for motorcyclists (16,678) to be 18% higher than that for car drivers (11,491). This occurs despite motorcyclists spending on average approximately three minutes less time traveling around the course. As noted in the previous chapter, a total of 8065 information objects were identified from within this total word count. Motorcyclists provided a total of 5022 information objects compared to 3043 for car drivers. This difference is significant according to a binomial test ($p < 0.01$).

The largest difference in total word count occurs in motorway driving and junctions, with motorcyclists providing 23% and 20.7% more verbal content respectively than car drivers. Controlling for the effect of each road section's mileage to produce a normalised 'words per mile' metric reveals a distinct pattern. Overall the fastest roads, with speed limits of 70 mph (i.e. motorways), 60 mph (i.e. major and country roads) and 40 mph (i.e. urban roads) produce less than 150 words per mile. Junctions and residential roads (with 30-mph limits) produce in excess of 350.

The first point to make is a methodological one. Clearly there is sufficient spare mental capacity, particularly for motorcyclists, for a rich verbal commentary to be provided across all road types. Indeed, the more challenging road types yield more content rather than less, which is what interference due to workload might otherwise suggest. The second point is a theoretical one. It is evident motorcyclists are able to extract more semantic content from the same situations than car drivers. Furthermore, it would seem the quantity of semantic content is contingent on the speed and hazard incidence rate of different road types.

Hazard incidence rate is a concept used in police driver training. A hazard is defined by Mares et al. (2013) as anything potentially dangerous with the potential to cause a change in vehicle position or speed. Clearly, some road types such as motorways, with restricted access, grade separated junctions, lower speed differentials, and gentle alignments have a lower hazard incidence rate than a busy urban road, with unrestricted access, at-grade crossings and potentially unfavourable geometry. In other words, 30 mph in an urban setting typically provides many more hazards per mile than 70 mph on a motorway. Differences in word count, therefore, seem to reflect different hazard incidence rates.

7.3.2 Propositional Networks: Analysis of Structure

A total of 12 propositional networks were produced in Leximancer[TM] from the semantic content captured in the verbal transcripts, six for each of the two road user types (motorcyclists and car drivers). These six networks refer in turn to the six road types encountered around the test route (motorway, major, country, urban, residential roads and junctions). Analysis of these networks now proceeds on the basis of their structure. The structural analysis employs techniques from graph theory to view the semantic networks in terms of nodes (n) and edges (e). These procedures help to reveal important underlying structural properties of the semantic networks which are not readily apparent from visual inspection alone. The metrics used are density, diameter and centrality.

Density is given by the formula:

$$\text{Network Density} = \frac{2e}{n(n-1)}$$

where e represents the number of edges or links in the semantic network and n is the number of nodes or semantic concepts. The value of network density ranges from 0 (no concepts connected to any other concepts) to 1 (every concept connected to every other concept) (Kakimoto et al., 2006). Density is a metric which refers to the propositional network as a whole and is a measure of its overall level of interconnectivity. Higher levels of interconnectivity suggest a richer set of semantic links and a well-integrated set of concepts. A denser network is also likely to have more well-connected concepts and shorter average path lengths. In order to diagnose the latter, a further metric is employed: diameter.

Diameter is given by the formula:

$$\text{Diameter} = \max_{\text{uy}} d(n_i, n_j)$$

where $d(n_i, n_j)$ is the largest number of concepts in the network which have to be traversed in order to get from one concept to another. Diameter, like density, is another metric which refers to the network as a whole. Generally speaking, the bigger the diameter the more concepts within the propositional network exist on a particular route through it. Again, generally speaking, a denser network will have a smaller diameter because the routes across it are shorter and more direct. A less dense network will have a larger diameter because routes across it have to traverse a number of intervening semantic concepts. This measure is related to the idea of clustering and to individual semantic concepts which are more or less well connected than others. In order to further diagnose this aspect of network structure yet another metric can be deployed: centrality.

Centrality is given by the formula:

$$\text{Centrality} = \frac{\sum_{i=1; j=1}^{g} \delta_{ij}}{\sum_{j=1}^{g} (\delta_{ij} + \delta_{ji})}$$

where g is the number of concepts in the semantic network (its size) and δ_{ji} is the number of edges (e) on the shortest path between concepts i and j (or geodesic distance) (Houghton et al., 2006). Centrality gives an indication of the prominence each concept has within the semantic network. Concepts with high centrality have, on average, a short distance (measured in edges) to other concepts and are likely to be well clustered and near the centre of the network. Concepts with low centrality are likely to be on the periphery of the network and to be semantically distant from other concepts.

Figure 7.2 presents the results of applying density, diameter and centrality to the 12 propositional networks created from the verbal protocol data. The radar plots show the structural metrics have some contingency upon road and road user type. In other words, the structure of driver's and rider's SA varies.

The mean level of interconnectedness of the propositional networks, as measured using the density metric, is 0.07 for car drivers and 0.08 for motorcyclists. This difference is very small with an almost identical

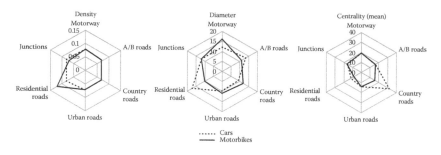

Figure 7.2 Radar plots of network density, diameter and centrality, and their contingency on road and road user type.

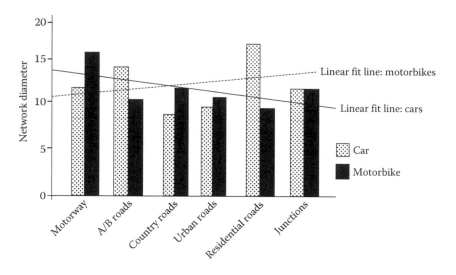

Figure 7.3 Results of diameter versus road type. Linear fit lines show that as road speeds and hazard incidence rates increase, the diameter of the semantic networks increases for car drivers and decreases for motorcyclists.

level of density across most road types. However, it can be observed how the semantic networks for motorcyclists become more densely interconnected when travelling over residential roads (0.12 compared to 0.08).

The results for diameter show that while the overall level of interconnectedness is broadly similar across road user types, the detail of that interconnectedness does differ. Figure 7.3 reveals an interesting pattern. A linear trend line seems to suggest that as road speeds, and hazard incidence rates, increase, the diameter of the semantic networks for motorcyclists decreases. This means the extent of direct access to semantic concepts increases with hazard incidence rate. The reverse trend seems true for car drivers.

Analysis of the metric centrality shows that, overall, as road speeds decrease so too does the mean level of clustering. In other words, as speeds increase specific semantic concepts become much more relevant than others. An exception to this overall pattern is when travelling over country roads, where the average level of clustering increases markedly for car drivers. A less dramatic increase can also be observed for both road users at junctions, where the mean level of clustering increases once more.

In summary, the overall level of semantic interconnectivity is broadly comparable between motorcyclists and car drivers. The main finding, however, is that while word counts increase with hazard incidence rate, the prominence of individual concepts within the DSA network tends to decrease for both road users. This seems to reflect the presence of a broader range of hazards (e.g. parked cars, hidden turnings, pedestrians etc.) for slower urban and residential roads compared to motorways and major roads, where none of the above hazards apply and attention is focused on a much reduced, but nonetheless important subset. Another key structural difference between motorcyclists and car drivers seems to be with respect to diameter. Here, average path lengths between semantic concepts decrease with hazard incident rate for motorcyclists (suggesting a more integrated mental representation), and increase for car drivers (suggesting a less integrated mental representation).

7.3.3 Propositional Networks: Phenotype and Genotype Schemata

An important observation is that despite quite high levels of overall structural similarity between semantic networks, there is a high level of dissimilarity in content. Less than half (mean = 43.3%) of individual semantic concepts for motorcyclists are shared with car drivers. This once again brings into sharp relief the differences between shared and compatible SA (e.g. Endsley and Robertson, 2000; Shu and Furuta, 2005; Salmon et al., 2009). The fact that a high degree of similarity exists in the structure of the networks despite less than half the concepts being shared prompts two further levels of analysis: a structural examination and a more detailed analysis of shared content.

The structural analysis proceeds by taking the 43.3% of shared semantic concepts and presenting them as a set of 30 individual concepts which are common across all road types and road users. From these a master semantic network is created. The 30 shared concepts represent the nodes for this network (n) while the interconnections between them (the edges, e) are based on those contained in the individual networks. For example, concept A might be linked to concept B in two individual networks (e.g. the networks for motorways and major roads). This in turn creates an edge value of 2 in the master network. The master network, therefore, represents the totality of shared concepts extracted from the verbal protocol data and the totality of links within the separate networks that join them. Figure 7.4 shows this master network of shared semantic concepts in relation to the differing patterns of activation when the data is partitioned into car drivers and motorcyclists.

Under the null hypothesis, it would be expected that the interconnections between shared semantic concepts for car drivers and motorcyclists would be the same. This is not the case. From Figure 7.4 it is visually apparent how the structure differs; moreover, the number of edges is slightly greater in the motorcycle network (e = 36) compared to the car network (e = 33). As a result, there is a small difference in overall interconnection (i.e. density) of 0.076 for car drivers and 0.083 for motorcyclists.

Slightly more pronounced are the overall differences in network diameter. Here it can be noted how the propositional network of shared concepts for motorcyclists (diameter = 5) is smaller than the

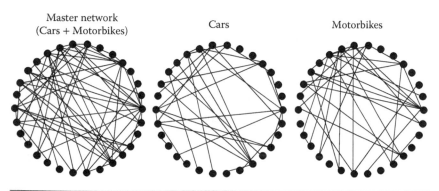

Master network
(Cars + Motorbikes)   Cars   Motorbikes

Figure 7.4 Master network created from shared concepts, with links between concepts derived from individual semantic networks.

same network for car drivers (diameter = 7). As before, this indicates motorcyclists have slightly more direct access, without intervening concepts on pathways across the network, to the core set of concepts common to both types of road user.

The most pronounced results refer to centrality. It can be observed how 63% of shared concepts increase their centrality/clustering in the motorcycle network compared to the car network. 6.7% of concepts remain unchanged (the null hypothesis) while 30% of the shared concepts decrease in the car network compared to the motorcycle network. More than double the number of shared concepts are more clustered, and therefore more closely linked semantically, again in the motorcycle network compared to the car network.

Overall, these findings communicate that whilst 30 individual semantic concepts are identical between car drivers and motorcyclists, their interconnection differs markedly. Structure is independent of content. Table 7.1 provides a further interpretation of the findings. The mean difference in centrality between motorcyclists and car drivers is observed to be 0.69 (SD = 6.35). This means 19 of these shared objects differ between road user types by less than one standard deviation either side of the mean. These semantic concepts therefore remain relatively enduring regardless of road user and can thus be related to the DSA concept of Genotype Schemata.

The five semantic concepts which increase their centrality beyond one standard deviation of the mean for motorcyclists, and the four semantic concepts which do the same for car drivers, can be associated with Phenotype Schemata. Because their importance changes meaningfully depending on road user type, they can be understood as relating to local state-specific schema brought to bear on a specific situation. Table 7.1 shows how the shared information objects break down into Genotype and Phenotype Schemata based on this criterion.

From Table 7.1 it can be seen how Genotype Schemata seem to relate to manoeuvring 'round' obstacles; overtaking; selection of 'gears'; awareness of speed 'limits' and control of 'speed'; and understanding the 'traffic' situation and 'road' layouts, such as 'roundabouts', 'lanes', traffic 'lights' and 'corners'. They also relate to events that are 'coming' or that are happening 'ahead' or those which are occurring 'behind' and observable in 'mirrors'. They also seem to refer to actions

Table 7.1   Concepts Whose Difference Value is within 1 SD of the Mean Are Defined as Genotype Schemata.

CONCEPT	EXAMPLE	DIFFERENCE (% RELEVANCE)	SD
Noticing	"noticing a yellow temporary sign for diverted traffic"	12.22	
Checking	"checking all round as we move away"	11.28	
Moving	"bit of traffic turning in front of us so we're not moving"	10.94	Phenotype Schemata
Doing	"we're doing well under 40 that's for sure"	9.4	
Check	"need to turn right, check both mirrors and indicate"	8.24	
Round	"looking at the big truck, [he's] not signalling, not going to argue with him, he is coming round"	4.46	
Gear	"…just kick it up a gear again, about 3000 rpm"	3.35	
Speed	"picking up the speed just very gently"	3.29	
Limit	"ok, we're entering a 30 [speed] limit"	2.99	
Traffic	"joining the M25 now…lots of traffic braking and switching lanes"	2.7	
Roundabout	"gently roll off, I can let everything pass me, on the roundabout, and over we go"	2.5	
Lane	"I'll just pull over into this left hand lane"	2.43	
Slow	"just let the car slow itself down as I indicate"	1.99	Genotype Schemata Mean = 0.7 +/− 1 × SD (6.35)
Car	"oh, just changed into 2nd at bit too higher revs there so the car jolted"	1.2	
Lights	"…and the lights are red"	0.72	
Take	"unfamiliar junction here so I'll take it easy"	0.45	
Coming	"nobody coming the opposite way"	0.44	
Ahead	"traffic building up ahead"	0.28	
Mirrors	"just going to check the mirrors to make sure it's ok to pull out…"	0.14	
Bend	"going to slow down for this bend"	0	
Braking	"easing off, braking a bit more…"	0	
Corner	"that's fine, this corner is slightly blind…"	−1.12	
Behind	"check the car behind"	−2.04	
Going	"just going to pull in and let them through"	−2.14	
Road	"…and road conditions look very dirty actually, lot of mud, there's obviously some road works going on?"	−3.98	

(*Continued*)

Table 7.1 (*Continued*) Concepts Whose Difference Value is within 1 SD of the Mean Are Defined as Genotype Schemata.

CONCEPT	EXAMPLE	DIFFERENCE (% RELEVANCE)	SD
Pulling	"no cars coming, pulling forward..."	−4.04	
Hill	"use plenty of revs to it up this hill"	−9.1	
Look	"some kids by the parked cars to look out for"	−9.95	Phenotype Schemata
Cars	"potential for cars pulling out in front of me"	−12.64	
Front	"clear on the road in front"	−13.09	

Note: Concepts whose difference value is outside 1 SD are regarded as Phenotype Schemata.

that are happening currently (such as 'pulling away' and 'pulling off') and actions which are 'going' to happen in the future.

Phenotype Schemata, or local state-specific routines, for motorcyclists include a lot of 'noticing', 'checking', 'doing' and 'moving' whereas for car drivers they involve 'looking', other 'cars', events specifically happening in 'front' and awareness of a particularly steep 'hill' (which does not affect the motorcyclists to any great degree due to higher power to weight ratios). The main differences between motorcyclists and car drivers, at this simple level of analysis, seem to be in the process of acquiring situation awareness, or so-called situational assessment. Motorcyclists seem to place greater emphasis on activities concerned with the attainment of SA (e.g. noticing and checking) whereas car drivers seem to refer to particular sources of information (e.g. other cars and events in front).

7.3.4 Propositional Networks: Thematic Analysis

In Leximancer™ concept groupings are referred to as 'themes'. These help to raise the level of analysis from the individual items of sometimes rather idiosyncratic keywords to that of broader, highly connected clusters related to how a situation is interpreted. Themes are ascribed a relevance value by Leximancer™. This is derived from the number of times the theme occurs as a proportion of the most frequently occurring concept (Smith, 2003).

There are a total of 64 individual themes extracted from the 12 semantic networks. Not all of these themes score highly in terms of relevance. The data is, therefore, filtered to capture those scoring in excess of 70% within either (or both) of the motorcyclist and car driver data sets. The filtering process reduces the number of themes from 64 to a high scoring subset of 20. Table 7.2 presents a summary of the results obtained. Under the null hypothesis it would once again be expected that the matrix of populated cells in Table 7.2, and the relevance values they contain, would be the same for motorcyclists and car drivers (a difference value of zero). Once again, this is not the case. Visual inspection of Table 7.2 reveals differences between road users. Out of the 120 cells contained in the matrix, 49 are not equal to zero. Of those 49, 21 show differences in relevance of 70% or more. Of those 21, 11 have increased relevance for motorcyclists and 10 for car drivers. There is 59.2% thematic overlap between motorcyclists and car drivers but, by the same token, 41.8% of strong thematic difference. This overall difference continues into specific road types. For motorcyclists, the pattern of results is consistent with the earlier findings on network diameter and centrality. Generally speaking, as road speed decreases and the hazard incidence rate increases, the number of themes, and their relevance, tends to drop. The following sections review the findings in more detail:

Motorway: Car drivers interpreted this situation in relation to events behind and in front of them. They also interpreted it in terms of the road itself, looking at the configuration of lanes and shoulders, in particular, their position on it and the relative position of others. Motorcyclists interpret this situation quite differently. Much more relevant is what is 'ahead', what is 'coming' towards them, and what is in 'front', including the slip road they plan to take in order to leave the motorway. Motorcycles are much more manoeuvrable so are not as constrained to lanes as car drivers, and more responsive longitudinally even at high speeds, meaning they can accelerate very rapidly. As such, events behind tend to be receding rather than encroaching, compared to events in front, which are approaching and need to be dealt with. Based on this reading of the propositional network, and despite thematic differences, there would nonetheless seem to be evidence of cognitive compatibility between car drivers and motorcyclists: the former engage with this situation with schema relevant to events behind them, whilst motorcyclists bring

Table 7.2 Matrix of Themes and Road Types

	MOTOR-WAY	MAJOR ROAD	COUNTRY ROAD	URBAN ROAD	RESIDENTIAL ROAD	JUNCTION
Ahead	96	−30	12	0	0	0
Behind	−16	−69	0	−14	−48	0
Car	0	0	74	0	54	0
Cars	0	0	0	0	−77	0
Coming	13	46	−14	100	−100	10
Corner	0	0	−94	0	0	0
Front	−48	84	24	0	0	−12
Gear	0	0	0	78	12	0
Going	27	0	−83	0	−60	7
Lane	−43	54	0	0	0	−84
Moving	78	0	0	0	0	0
Pull	0	0	0	−100	0	−12
Road	−67	100	100	−23	0	−80
Shoulder	−36	0	0	0	0	0
Signal	0	0	0	0	0	70
Slow	0	0	−71	0	0	0
Third	0	0	−86	−4	0	−56
Traffic	31	0	0	0	0	100
Van	0	0	0	0	0	−80
Slip	100	0	0	0	0	0
Number of themes increasing in relevance for motorcyclists	6 (Sum 480%)	4 (Sum 284%)	4 (Sum 210%)	2 (Sum 178%)	1 (Sum 54%)	4 (Sum 187%)
Number of themes increasing in relevance for car drivers	5 (Sum 210%)	2 (Sum 99%)	5 (Sum 348%)	4 (Sum 141%)	4 (Sum 285%)	4 (Sum 324%)
Number of themes remaining the same for both road users	9 (45% Overlap)	14 (70% Overlap)	11 (55% Overlap)	14 (70% Overlap)	10 (50% Overlap)	14 (70% Overlap)

Note: Cells populated with the difference in % relevance between motorcyclists (positive numbers), car drivers (negative numbers) and instances of thematic overlap (zero values/null hypothesis).

to bear schema relevant to what is in front. On motorways at least, motorcyclist and car driver SA seems to be cognitively compatible, in certain important respects, with different awareness working in a compatible way.

Major Road: With freedom of movement becoming more constrained due to increasing traffic movements, fewer lanes, and narrower road widths, the semantic networks for motorcyclists now become more similar to the semantic networks for car drivers on motorways. The situation is now interpreted more in terms of what is 'behind', the 'road' conditions in terms of braking and manoeuvring and 'lanes'. Car drivers, meanwhile, continue to interpret this situation in terms of what is behind, but now also understand it in terms of traffic 'coming' from other directions. Again, cognitive compatibility is in evidence with both road users interpreting the situation in terms of 'lanes'. Motorcyclists now bring to bear schema related to encroachments from behind whilst car drivers deploy schema related to events 'ahead'.

Country Road: The semantic networks for car drivers now reflects a situational interpretation which has switched from 'behind' to 'ahead'. The challenge in these situations is less to do with traffic conflicts, lanes, and road layout and more to do negotiating natural features and corners. Planning ahead also rises to prominence. Motorcyclists also interpret this situation in terms of what is ahead but, surprisingly, give less direct emphasis to hazards such as corners. That being said, much more indirect emphasis is made in relation to the state of the road itself in terms of adhesion, conditions etc. Unlike motorways and country roads, areas of cognitive incompatibility start to be revealed. For example, both road users now bring to bear schema related to 'ahead'. For car drivers the concept of 'ahead' occurs in conjunction with schema about planned action. For motorcyclists the concept of 'ahead' occurs in conjunction with the concept of 'car' – specifically, what 'cars' are actually doing. While the concept of 'ahead' is shared between these two road users, the way in which it is connected to other concepts is quite different. In essence, the two mental theories of the situation have become uncoupled. The motorcyclists operate with anticipatory schema incorporating awareness of the car in front, and of planned actions occurring thereof; the car driver's schema is to focus on what is ahead, not approaching from behind.

Urban Road: Both motorcyclists and car drivers interpret this situation in terms of events 'behind'. Once again, much greater emphasis is placed by motorcyclists on what is 'ahead'. Selection of the appropriate 'gear' is also important for both road users. The prominence of the concept 'slow' for car drivers reflects the fact that progress is much more frequently impeded compared to motorcyclists. They are able to use their awareness of what is 'ahead', combined with the characteristics of their vehicle, to maintain progress. Evidence of cognitive incompatibility is once again indicated, with anticipatory schema of traffic that is 'coming' only in evidence for motorcyclists. Car drivers seem to be operating in the realm of schemas about events behind, and how this interacts with correct road positioning and lane selection.

Residential Road: In this situation, both road users refer to individual vehicles around them, but they are of much more prominence to car drivers whose progress is once again more significantly impeded. Lower speeds also raise to prominence the concept of being in the correct 'gear' for motorcyclists. In this situation, there is little to suggest significant cognitive incompatibility. Motorcyclists seem to be relatively unaffected by the impediments that seem to drive much of the active schema for car drivers. Instead, their schema seems more related to the behaviour of cars around them, and the proper control of their machine.

Junctions: This situation is interpreted very differently between the two road users. This is particularly interesting given that a major source of traffic conflicts between cars and motorcycles occurs at junctions. Car drivers are once again concerned with the appropriate travel 'lane', 'pulling' away and specific vehicles (e.g. 'van'). The semantic networks for motorcyclists reflect a need to give information to other road users in the form of 'signals' and a wider appreciation of 'traffic' (rather than specific vehicles). In this situation there are areas of cognitive compatibility, with motorcyclists clearly bringing to bear schema which attempt to compensate for car drivers, who are more preoccupied with road position and lane manoeuvres. Cognitive incompatibility is in evidence because car drivers seem to be operating with schema concerning events 'coming' or in 'front', with no corresponding relevance given to events 'behind'. As a result, the effect of signals and other measures motorcyclists take to increase awareness may be limited.

7.4 Summary

This chapter has built on the last, showing how a reliable, automated, semantic network creation process, coupled to concurrent verbal protocol data, can take the notion of systemic SA and push it further. This enables questions about how different road users experience the same road situation to be explored.

SA-based insights are provided into the key transport safety challenge of compatibility between different road users. From this analysis it is clear motorcyclists interpret the same road situations differently to car drivers. In many road circumstances this interpretation appears to have important areas of mutual reinforcement, with strong stereotypical responses which favour the anticipation of each other. This is not the case for all road types.

Not surprisingly, the two road types of most concern to motorcycle safety (i.e. junctions and country roads) are interpreted differently and in ways that are more difficult to reconcile. The study thus shows a good degree of triangulation with acknowledged sites of car/motorcycle incompatibility such as junctions (i.e. 'right of way' accidents) and country roads (i.e. single vehicle accidents on bends). The exploratory analysis is also compatible with a number of more practical countermeasures, all of which present themselves as candidates for further in-depth study into how potential gulfs of evaluation can be narrowed.

The first of these is driver and rider training. The present analysis method can help to equip drivers with a form of 'meta-awareness' of their own propensity towards certain cognitive states in certain situations. Driver training could, for example, provide coaching and tuition on the need to conduct regular rearward checks on country roads and at junctions, as faster vehicles may be approaching from behind.

The second countermeasure refers to infrastructure. It could involve road signs that do not warn of events ahead, but instead warn of potential events behind (e.g. 'faster traffic approaching behind', 'check mirrors now' etc.). Taking this even further raises the possibility of deploying a concept from the railway industry called 'route drivability'

(Hamilton et al., 2007). In essence, this is a form of 'analytical proto-typing' in which proposed changes in routes, signalling, signage etc. are tested in terms of driver workload. For road transport, the verbal protocol/semantic network method (in cooperation with others) could serve a similar analytical prototyping purpose. The method outputs could be used to ascertain how road situations are interpreted, how physical features could be used to modify that interpretation in favour-able ways, and to define cognitively the optimum type and placement of road signs. This, in turn, links very well to a growing body of work in the Self-Explaining Roads (SER) domain (e.g. Walker et al., 2013).

A third intervention, linked in some ways to the first, is the con-cept of cross-mode training. For road transport there is distinction to be made between specific vehicle control skills (e.g. clutch control, hill starts, reversing etc.) and mode-independent skills (e.g. road and traffic awareness, giving indications, rights of way etc.), or, in other words, the 'basic' and 'operational' driving tasks and the 'tactical' and 'strategic' tasks specified in the earlier Hierarchical Task Analysis of Driving (HTAoD – see Chapter 4 and the Appendix). Advanced driver training offers just such an approach, and its role in enhancing driver SA is explored next. For now, though, it is clear just what a use-ful experimental manipulation the examination of motorcyclists is. In vehicle feedback terms they are vastly different to cars and this does indeed affect SA. To quote *Zen and the Art of Motorcycle Maintenance* once more: "Unless you're fond of hollering you don't make great con-versation on a running cycle. Instead you spend your time being aware of things and meditating on them" (Pirsig, 1974, p. 17). Except per-haps if you are following an experimental participant in an offset road position while they attack the experimental route as if it were the Isle of Man TT.

8

DRIVER TRAINING AND
SITUATION AWARENESS

8.1 Introduction

8.1.1 *The Normative Gold Standard*

In 2005 we had the pleasure of being involved in a project with the UK Institute of Advanced Motorists (IAM). The connection was made, as it happens, via the extensive recruitment campaign embarked upon for the study described in Chapter 5. For us it opened up a fascinating insight into the world of highly trained police drivers. This included a visit to the Metropolitan Police Force's driver training facility located at the Peel Centre in North London, more commonly referred to by officers as simply 'Hendon'. Prior to this visit none of the authors had travelled at 130 mph around the M25 London Orbital Motorway, nor witnessed the flash of so many speed cameras being triggered. We travelled in convoy in unmarked BMW 330d's with lights and sirens active. Sitting in the passenger seat of the lead vehicle one had to resist an overriding temptation to tell the driver to pull over because there was a police car behind.

What struck us during these demonstration drives was their complete and utter safety. Far from the chaotic high-speed police chases of Hollywood film lore these were exceptionally smooth, measured and, most noticeably, not fast all of the time. At one moment we would be travelling at 85 mph along the middle of a busy London thoroughfare, sirens blaring; the next we would be rolling at literally 3 mph as we approached a pedestrian crossing. Extreme courtesy was in evidence: the trained pursuit drivers would give abundant information to other road users. Indicators were used. Horns sounded. Waves of thanks were issued to other drivers. Eye contact was established liberally. In addition to the information given to other road users it was also taken. Times of day were noted in relation to school traffic, brake lights were noticed and pedestrian movements observed. The extent and diversity of forward planning was remarkable. The vehicle was settled at all times. A large and noticeable part of this was road positioning, with the need to collect as much information from the forward scene necessitating wide variations in where the car was placed. It was clear this, as much as anything, was enabling the drivers to create far better situations to be subsequently aware of. At no point was a situation arrived in that required sudden changes in speed or trajectory. There were also no elaborate power-slides, shootouts, jumps, collisions or piles of cardboard boxes to drive through.

What we observed, in fact, was an extreme demonstration of the Hierarchical Task Analysis of driving described earlier. This was the normative 'gold standard' in practice during which the true extent of driver situation awareness (SA) was rendered to us in a way that could be easily appreciated. This is because, for the duration of the drive, the drivers were providing a rapid and continuous verbal commentary, not for experimental purposes, but because it has been a part of police driver training for over 70 years. Numerous examples of this can be found on YouTube (e.g. www.youtube.com/watch?v=MRmiaQqWt7Y) and they make for fascinating viewing (Figure 8.1). The measured approach, despite the occasional high speeds, is due in large part to an underlying 'system' of car control which the IAM also uses as the foundation for their driver coaching. As noted in the previous chapter, this is a form of coaching which relates most closely to higher-level tactical and strategic driving tasks and, as such, is independent of vehicle type. IAM training is delivered to private car owners as it is to motorcyclists and heavy goods vehicle drivers. In experimental terms advanced driver coaching enables a number of pertinent driver-SA questions to be further explored, and assumptions about the nature of SA further challenged. Once again it was an opportunity not to be missed.

If advanced driver coaching has an effect on SA then based on previous chapters it should:

- Increase the number of information elements in the driver's working memory
- Increase the interconnection between those elements
- Increase the amount of 'new' information in memory

- Increase the prominence of existing information
- Stimulate behaviours that help drivers evolve better situations to be aware of

The systemic, networked, DSA approach described in Chapters 6 and 7 is deployed again, this time within a longitudinal on-road study. The study involved three groups of 25 drivers, all of whom were measured pre- and post-intervention. One experimental group was subject to advanced driver coaching while two further groups provided control for time, and for being accompanied whilst driving. Empirical support was found for all five hypotheses. Advanced driving does improve driver SA but not necessarily in the way existing situation-focused, closed-loop models of the concept might predict.

8.1.2 Advanced Driving

According to the IAM, advanced driving is not only about how to handle a vehicle. The driver must be absorbed into the complete culture associated with this discipline, becoming a thinking driver. The underlying aim of advanced driving, therefore, is to develop the driver's skills in courtesy, restraint and spirit of co-operation with fellow road users, combined with an attitude of controlled professional and methodical driving, displaying a high level of concentration. The IAM share with many similar bodies a desire not to create an elite minority of experts. Their aim is to convert a high proportion of the driving community into 'thinking safe drivers' who are knowledgeable, enthusiastic and competent.

The observers who conduct advanced driver courses aim to produce a candidate for the advanced test who can display 'quiet efficiency'. The observer performs in a 'coaching' rather than 'training' role. They are not teaching someone how to drive a car. The candidate already has a driving licence. The observer's role is to refine their skills and introduce new concepts and more advanced skills to enable the associate to achieve a higher standard. In doing this the observer will be acting in the role of a coach, co-ordinator and adviser but not instructor. The observer will not only add new, or refine further, the associate's technical skills but will introduce and build the attitudes so important to being a successful 'advanced driver'.

The purpose of this chapter is to examine the effect of advanced driver training on SA. It should be apparent from the previous chapters that vehicle feedback, and by implication driver SA, is changing, and will change further in future. It is also apparent drivers seem able to compensate for these changes but in ways that are not fully understood. It seems likely, furthermore, that as increasing numbers of new vehicle types enter the traffic fleet new approaches to training will be required to help equip drivers to operate them. With the role of driver SA in behaviour revealed it seems natural to ask whether improved driver SA skills, knowledge and behaviour can, in fact, be trained. In answering this practical question it is also possible to dig deeper into the theoretical underpinnings of SA itself.

The term 'advanced driving' covers a wide range of interventions, some more 'advanced' than others. To be clear, the class of training forming the topic of this chapter is sometimes referred to as 'Tactical' or 'Defensive' driving. Crucially, the underlying techniques derive from police driver training and are based on an explicit 'system of car control' (Police Foundation, 2007; IAM, 2007). The system of car control can be defined as a structured, sequential, behavioural heuristic. It is a form of drill which enables broad classes of driving behaviour to be emitted in sequence so the driver can approach and negotiate hazards in a consistent manner, and in such a way as to reduce risk (Police Foundation, 2007; IAM, 2007). Hazards are defined as any entity or artefact encountered in the external road environment, specifically physical elements (such as junctions or bends), risks (or potential risks) arising from the position or movement of other vehicles, and problems arising from environmental conditions (road surface, weather and so on) (Police Foundation, 2007). Hazards require the driver to potentially change speed and/or direction. The system of car control underlying the IAM's training method breaks down the response to such hazards into five phases, as shown in Table 8.1. The first phase is called 'information' and has a very clear link with SA.

The system of car control relies on the driver performing different phases depending on the driving context, entering one stage of the system and exiting another, with the information phase overarching all. The information phase describes both a set of physical behaviours (i.e. activities associated with gathering information on the driving

Table 8.1 The System of Car Control

PHASE	DESCRIPTION
Phase 1: Information	"You always need to be seeking information to plan your driving and you should provide information whenever other road users could benefit from it" (Mares et al., 2013, p. 30)
Phase 2: Position	"Position yourself so that you can negotiate the hazard/s safely and smoothly" (Mares et al., 2013, p. 32)
Phase 3: Speed	"Adjust your speed to that appropriate for the hazard" (Mares et al., 2013, p. 33)
Phase 4: Gear	"Select the appropriate gear for the speed at which you intend to negotiate the hazard" (Mares et al., 2013, p. 33)
Phase 5: Accelerate	"Use the throttle to maintain speed and stability through the hazard [...] choose an appropriate point to accelerate safely and smoothly" (Mares et al., 2013, p. 33).

Source: Mares, P., Coyne, P. and MacDonald, B. *Roadcraft: The Police Driver's Handbook.* Stationary Office Books, London, 2013.

context) and 'cognitive behaviours' (i.e. perceiving, thinking and, in everything but name, SA). Phases two to five of the system are ostensibly 'doing' phases which relate more directly to car control. Phases two to five are linked together intelligently and in sequence by information the driver is extracting from their environment. It is important to stress that the system embodies a degree of flexibility. An important part of the training process is on the intelligent application of the system, for which highly functioning driver SA is required.

8.1.3 Predictive SA

We saw in the previous chapter how a better situational model is likely to be more parsimonious: the more parsimonious, the more the 'information in working memory' (e.g. Bell and Lyon, 2000) is forced into higher, more implicit levels of abstraction. The paradox arising from this is that the better the driver's 'situational model' of their situation the 'less' likely it is to look like the object or situation being perceived. This is a significant conceptual challenge for more conventional models of SA based on input–output processing models of cognition, as we will see later. One of the reasons why advanced driving is useful to study is because reflecting it off such theories shows that the people for whom a canonical situation model might literally apply are novice drivers, those couched at Rasmussen's skill-based level of procedural, declarative knowledge (Rasmussen et al., 1994), not expert

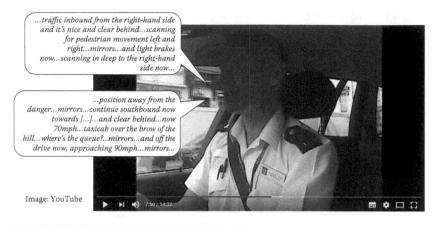

Figure 8.1   The system of car control as verbalised by a trained police driver. The role of information (and SA) is clear.

advanced drivers operating at the knowledge-based level. The interesting point about advanced driver training is that it aims to bring previously automatized processes back into consciousness, not least via the concurrent verbal protocol approach. This makes processes of expert decision making, like SA, more amenable to inspection (e.g. Summala, 2000).

Two exploratory hypotheses emerge from this. Hypothesis #1 is that advanced drivers should have quantitatively more situational information in working memory compared to similarly experienced drivers. This is because the coaching intervention is designed to bring such information into consciousness, not necessarily because their mental theory of the situation contains more information. Differences in the quantity of information elements are interesting to a degree, but perhaps less so than how they are connected together (as we saw earlier in Chapter 7). Hypothesis #2, therefore, would anticipate that regardless of the quantity of information, a greater extent of linkage would exist between situational information as a result of advanced driver training (e.g. Flach et al., 2004). This, of course, is entirely consistent with Ausubel's (1963) theory of learning which underpins the propositional networks embedded within the DSA approach.

As touched upon in earlier chapters, the current state of practice in SA carries with it a certain 'linear' flavour. This is an area of SA theory which is hotly contested (e.g. Endsley, 2015; Stanton et al., 2015). For us, evidence of this linearity can be seen in the implicit control-theory logic of sequential flows and feedback embodied in Endsley's popular model of SA (Endsley, 1995). Although disputed, with linear notions contested as a form of 'fallacy' promulgated in the literature in error, the fact such models feature Level '1', '2' and '3' SA does little to help disabuse users of the notion. Neither does the following, quoted from one of the most frequently cited texts in the field of ergonomics:

"The first step in achieving SA is to perceive the status, attributes, and dynamics of relevant elements in the environment" (Endsley, 1995, p. 36).

"Comprehension of the situation is based on the synthesis of disjointed level one elements" (p. 36).

"Based on knowledge of level one elements, particularly when put together to form patterns with the other elements (gestalt) the decision maker forms a holistic picture of the environment, comprehending the significance of objects and events" (p. 37).

"...the third and highest level of SA. This is achieved through knowledge of the status and dynamics of the elements and comprehension of the situation (both level 1 and level 2 SA)" (p. 37).

"A person can only achieve level 3 SA by having a good understanding of the situation (Level 2 SA) and the functioning and dynamics of the system they are working with" (Endsley and Jones, 2012, p. 18).

The more hotly contested nature of the theoretical debates in SA can be accessed from Volume 9 of the *Journal of Cognitive Engineering and Decision Making* and are a very worthwhile read. For now, though, whilst there can be little doubt a large proportion of driving performance does rely on an approximately linear cycle of input–processing–output–feedback, this need not be the case all the time. In fact, a larger portion of driving than is often acknowledged may be feed-forward. An extreme case of this is Driving Without Attention Mode (DWAM) (Kerr, 1991; May and Gale, 1998).

DWAM might be familiar to many readers. It is that "oh, I'm here!" feeling of arriving at a destination without clear recall of the specifics of the journey, particularly if that journey is highly familiar, like a well-trodden commute to work. Despite a lack of recall there must have been a functioning mental theory of the driving context, and concomitant SA, in order to have arrived safely. Kerr (1991) cites the automatisation of perception due to predictability in the environment as the cause for this. This explanation highlights the role of expectancy, or feed-forward control. In the language of SA DWAM appears to be a case of Level 1 SA (perception) no longer preceding Level 2 (comprehension) or 3 (projection). In other words, the first step in achieving SA is to save cognitive effort and rely on expectations about the status, attributes and dynamics of relevant elements in the environment, relying much more on 'knowledge in the head' than 'knowledge in the world'. DWAM seems to admit the possibility of non-linear awareness. To be fair, the dominant dialogue in current SA theories does acknowledge this, in theory if not always in practice.

From an advanced driving perspective the clue that similar non-linear processes are at work resides in the term 'defensive driving'. This brings forward notions of 'anticipation', which is to say instances where comprehension and projection occur before any overt 'situation focused' perception has occurred. One example is 'contingency plans', something advanced driver coaching encourages (Mares et al., 2013, p. 42). Contingency plans are analogous to Level 3 SA (projection) but they are developed without any overt Level 1 SA (perception). In other words, Level 3 SA is not driven from the extant realities of the situation in the strict feedback sense of the term, simply because nothing has been perceived yet to require it. Instead it is being driven from knowledge that is brought to a situation, 'expected', rehearsed through the extant mental theory of the situation, and made ready to use if triggered. It can be argued that something Level 1-like must have triggered this rehearsal process in the first place. Whatever it was, though, is detached from the immediate 'volume of time and space' that may or may not require it. To emphasise again, Endsley's dominant model of SA can provide an explanation for this to some extent, but it is far from the simple logic of Levels 1, 2 and 3. Indeed, it begins to reveal a form of intractable complexity as ever more loops, cognitive modules and pathways between them need to be invoked to explain what is happening, a form of complexity rarely confronted in the literature, and strongly suggestive of different models of SA being appropriate for different types of research question. We thus return to DSA and the interconnected network of information upon which different actors and agents will have distinct views and ownership.

Exploratory hypothesis #3, as a result of all this, states that not only would advanced drivers have more knowledge in working memory (Hypothesis #1), that it would be connected differently for advanced drivers compared to 'normal' drivers (Hypothesis #2), but that non-linear feed-forward awareness in the form of anticipation will require and/or generate new types of information. In addition, as a result of the advanced driver coaching, this information would be of greater importance/prominence in the situational model compared to non-advanced drivers.

8.1.4 Normative versus Formative SA

The current state of the art in SA also carries with it a 'normative' flavour, the tacit assumption being the objective situation provides a reference point for judging 'goodness' or 'badness' of SA (Endsley, 1988, p. 793). SA, however, is created for a purpose (Patrick and James, 2004). The outcomes of real-world advanced driver coaching seem to be better situations to be aware of rather than better awareness of merely average situations. The 25% reduction in post-test crashes following advanced driver training seems to testify to this (Hoinville et al., 1972).

Smith and Hancock (1995) have put it that SA can be viewed as "a generative process of knowledge creation" (p. 142) in which "[...] the environment informs the agent, modifying its knowledge. Knowledge directs the agent's activity in the environment. That activity samples and perhaps anticipates or alters the environment, which in turn informs the agent" (Smith and Hancock, 1995, p. 142) and so on in a cyclical manner. This raises further interesting questions about the linearity or otherwise of SA. Smith and Hancock refer to SA as 'constructive', which is to say the driver is a part of the situation they find themselves in and can influence its dynamics. This is made explicit in the advanced driving *Roadcraft* manual which states: "The aim [...] is to improve the skill and safety of your driving so that *you can make best use* [emphasis added] of road and traffic conditions [...]". Two further aspects of advanced driving are revealed by this.

Firstly let us consider the example of 'vanishing points' on bends (see Figure 8.2). Unlike Hypothesis #3, which talks of new knowledge, advanced driver training also enables existing information to

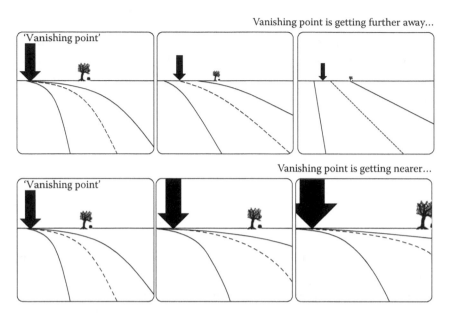

Figure 8.2 Advanced driver coaching enables drivers to take advantage of motion-parallax cues to help guide their speed choice in corners.

be 'used' in new ways. The use of vanishing points on bends enables the driver to use information that was always there (motion parallax cues) to now anticipate the curvature of the road ahead and adjust their speed as appropriate. Exploratory hypothesis #4, therefore, would predict not just an overall increase in the level of connectivity between information elements, but for certain individual elements shared between pre- and post-coaching to be elevated in status within the DSA network.

Secondly, the term 'use' also highlights the proactive nature of advanced driving and the ability to control the position and speed of the vehicle relative to everything else on the road. Indeed, "An accident or even a near miss usually represents a loss of this control" (Mares et al., 2013, p. xi). Framed in terms of Smith and Hancock's (1995) cyclical model of SA, information derived from sampling the environment during the information phase of the system 'directs activity'. The information phase 'directs' the position of the vehicle, its speed, the gear and use of acceleration, each phase being the cause of more information and more 're-direction' of behaviour (Figure 8.3).

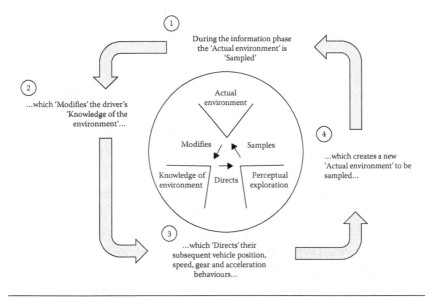

During the information phase
the 'Actual environment' is
'Sampled'

...which 'Modifies' the driver's
'Knowledge of the
environment'...

Actual
environment

Modifies   Samples

Knowledge of   Directs   Perceptual
environment   exploration

...which creates a new
'Actual environment' to be
sampled...

...which 'Directs' their
subsequent vehicle position,
speed, gear and acceleration
behaviours...

Figure 8.3   The system of car control mapped to the perceptual cycle model of cognition.

Thus, rather than linearly deterministic, in practice SA emerges as rather more evolutionary and probabilistic which, in turn, raises questions about judging 'goodness' or 'badness' of SA. Is good awareness of a bad situation the same as poor awareness of a good situation? If risks from hazards have been successfully minimised through the intelligent application of the system of car control, then does the more benign situation the driver has created result in fewer critical variables to be aware of? What does seem to be the case is that in evolving themselves towards better situations, advanced drivers should exhibit observably different driving behaviours compared to 'normal' drivers. This is exploratory hypothesis #5.

8.1.5 Measuring and Representing Driver SA

It is clear from the above that SA models and methods based on closed-loop processing and situation-focused assessment are appropriate for testing some of these hypotheses but not others. An approach responsive to these paradoxes, and one that can subject the five experimental hypotheses to a test, is again the DSA approach. In order to argue this case the individual SA lens will at points be slid across the frame to reveal just how different the results can be.

8.2 What Was Done

8.2.1 Design

The study performed in this chapter was designed to analyse the effect of advanced driver coaching on SA. Five key variables are subject to test: quantity of information elements in the DSA network (Hypothesis #1); interconnectivity of elements (Hypothesis #2); differences in the relative importance of elements (Hypotheses #3 and #4); and driving behaviours (Hypothesis #5). Data for hypotheses #1 to #4 are gathered using the Critical Decision Method (CDM) interview (O'Hare et al., 2000; Klein et al., 1989). This is an alternative to using verbal protocols as a way of constructing DSA networks. Hypothesis #5 relies on behavioural observations conducted during and after the assessed drives using a 12- and 26-item, respectively, check sheet based on standard IAM assessment materials.

The between-subjects factor was coaching intervention, which had three levels:

Group 1 was subject to full IAM coaching.
Group 2 was subject to 'observation' but no coaching or feedback.
Group 3 was neither observed nor coached.

This design provides one experimental group (Group 1 who had full IAM coaching) and two control groups (Group 2 controlled for the effects of simply being observed and Group 3 controlled for the effects of time).

The within-subjects factor was measurement interval, again with two levels. Drivers were measured using the CDM interview and direct behavioural observations pre-intervention (Time 1) and post-intervention (Time 2).

The three groups of drivers were matched on age, gender, experience and annual mileage. The different assessors all received standard training on how to complete the assessment material and their performance was further re-checked during the study. The coaching intervention for Groups 1 and 2 also took place using standardised on-road routes defined beforehand by the IAM.

8.2.2 Participants

75 drivers took part in the study. There were 21 female drivers and 54 males divided equally among the three experimental groups (each group had 25 drivers). Once again, despite efforts to the contrary, a gender-balanced sample was not achieved. To some extent this was outside the study team's control due to the opportunistic nature of the sampling, which relied on matching participants to whomever presented themselves for IAM coaching. Those that did tended to be male.

The drivers were aged between 23 and 65 years (mean of 44 years), drove on average 13,000 miles per year (minimum of 1500, maximum of 35,000), and had a mean of 23.8 years' experience (minimum 3.7, maximum 44). The three experimental groups were matched to within 0.7 years on age ($F(2,71) = 0.037$, $p = $ ns) and experience ($F(2,71) = 0.03$, $p = $ ns) and to within 1631 miles for annual mileage ($F(2,71) = 0.43$, $p = $ ns). The study granted in excess of 80% probability for detecting large effect sizes (or greater) on these matching variables if such effects were present. Evidently they were not and in all cases the obtained effect sizes were exceedingly small (partial eta squared = 0.001, 0.001 and 0.012 respectively). This indicates, statistically, that very close participant matching was achieved.

The study took place in the South West region of London and involved the participation of five IAM regional groups (South of London, Central London, North West London and Chilterns, Thames Valley, and Guilford and District). The two control groups were members of the public, also recruited from the SW region of London. Drivers in the control groups who were not presenting themselves for IAM training were paid between £75 and £300 for participating in the study, which took place over eight weeks and involved a commitment of between two and eight 45-minute drives.

8.2.3 Materials

8.2.3.1 Experimental Booklet A response booklet was constructed enabling drivers to be assessed in a standardised manner pre- and post-intervention (at Time 1 and Time 2 of the study). The response booklet contained a demographics questionnaire used to 'match' participants across groups, an on-line behavioural checklist, an off-line assessment sheet (both shown in Table 8.2), and a copy of the (off-line) CDM probe questions with spaces for responses (Table 8.3).

The on-line behavioural checklist was derived from the IAM's standard assessment materials and was completed during the assessed drives by the IAM observer. The off-line assessment sheet was also based on standard IAM materials and completed in the normal manner after the assessed drive was complete (again by the IAM observer). The CDM was also completed after the assessed drive in a supported self-report manner. The driver provided written responses

Table 8.2 On- and Off-Line Behavioural Checklist and Assessment Sheet

ON-LINE BEHAVIOURAL ASSESSMENTS (FREQUENCY OF APPROPRIATE/INAPPROPRIATE BEHAVIOURS NOTED)	OFF-LINE ASSESSMENT OF DRIVING BEHAVIOUR (1 = EXCELLENT, 5 = POOR)
Speed	Acceleration
Limit points approaching bends	Acceleration Sense
	Braking
	Clutch Control
System of car control	Gear Changing
	Use of Gears
Use of handbrake	Steering
Smoothness	Manoeuvring
Signalling	Concentration
Steering (correct grip and action)	Observation
	Anticipation
Positioning (for hazards)	Hazard Assessment
	Hazard Management
Headway	Road Position
Response to hazards	Speed Limits
	Overtaking
	Safe Progress
Use of mirrors	Mirrors
	Signalling
Gear changes	(Road) Surfaces
	Smoothness
	(Mechanical) Sympathy
	Courtesy
	Aptitude
	Commentary
	Knowledge

Table 8.3 CDM Probes

COGNITIVE CUE	SAMPLE QUESTION
Goal specification	What were your specific goals at the various decision points?
Goal identification	What features were you looking at when your formulated your decision?
	How did you know that you needed to make the decision?
	How did you know when to make the decision?
Expectancy	Describe how this affected your decision-making process.
Conceptual model	Are there situations in which your decision would have turned out differently?
	Describe the nature of these situations and the characteristics that would have changed the outcome of your decision.
Influence of uncertainty	At any stage, were you uncertain about either the reliability or the relevance of information that you had available?
	At any stage, were you uncertain about the appropriateness of the decision?
Information integration	What was the most important piece of information that you used to formulate the decision?
Situation awareness	What information did you have available to you when formulating the decision?
Situation assessment	Did you use all the information available to you when formulating the decision?
	Was there any additional information that you might have used to assist in the formulation the decision?
Options	Were there any other alternatives available to you other than the decision that you made?
	Why were these alternatives considered inappropriate?
	Was there any stage during the decision making process in which you found it difficult to process and integrate the information available?
Decision blocking	Describe precisely the nature of the situation.
Basis of choice	Do you think that you could develop a rule, based on your experience, which could assist another person to make the same decision successfully?
	Do you think that anyone else would be able to use this rule successfully?
	Why / Why not?
Generalisation	Were you at any time reminded of previous experiences in which a similar decision was made?
	Were you at any time reminded of previous experiences in which a different decision was made?

Source: O'Hare, D. et al., In J. Annett and N. A. Stanton [eds.], *Task Analysis* (pp. 170–190), Taylor and Francis, London, 2000.

to the cognitive probes in their own time with the observer on hand to facilitate as required.

8.2.3.2 Test Route All driving took place on public roads, in the driver's own vehicle, and within the south-west area of London. The test routes, though not identical, conform to standard guidelines laid down by the IAM. This means a full range of road types was featured, such as motorways, urban and rural roads, ensuring that all posted speed limits could be attained. All routes were approximately 25 miles in length leading to a driving time of approximately 45 minutes.

8.2.4 Procedure

8.2.4.1 Development of the Participant Pool Drivers who registered with the IAM for advanced driver coaching were approached by the observers from the appropriate London regional groups. In the course of the normal formalities they were asked whether they would like to volunteer for the study. Over the eight-week duration of the course the volunteers were given IAM coaching but had their SA and driving behaviour(s) measured at the beginning and end of the coaching period. As soon as the first completed response booklet was sent to the study team for analysis, the control group(s) were matched, selected, then contacted. Drivers for the control groups were also recruited from the south-west area of London. Demographic questionnaires were sent out to candidate drivers to establish their age, gender and annual driving mileage. This data assisted in matching them to the IAM group.

8.2.4.2 Procedure for the Advanced Driver Group The London IAM regional groups were responsible for conducting the observations and CDM interviews with the participants (called 'associates' by the IAM) who wished to take part in the study. Each associate was assessed at 'Time 1' and 'Time 2'. Time 1 was the initial start point of the advanced driver coaching programme and Time 2 was determined to be eight weeks hence, at which point the observers typically feel the associate to be ready to take the advanced test. Even if not, it represents a sufficient period for improvements to emerge. A capable associate would normally be 'test ready' (or close) when measured at

Time 2. After the response booklet was completed, the IAM observers would send it back to the study team for analysis.

8.2.4.3 Procedure for the Control groups Upon receipt of the first response booklet from the IAM the study team would select a matching control participant from the participant pool. Time 1 was determined to be the start point of the '8 drive group' (the group who were merely observed as they drove) or the '2 drive group' (the group for whom time was merely allowed to elapse between measurement at Times 1 and 2). Time 2 represented the end point of the observed or 'time elapsed' period. Each participant in the control group was assessed over a period of eight weeks. As noted above, this was determined to be the average time IAM associates took to be coached to test-ready status.

To reiterate, the first control group (2 drives) was assessed at Time 1 and then, after 8 weeks, was assessed at Time 2. They were neither coached in advanced driving techniques neither were they observed. The second control group (8 drives) was assessed at Time 1, then they were accompanied for 8 drives (1 drive per week over 8 weeks) then assessed at Time 2 (the final eighth drive). The first and eighth trial, during which drivers were observed and measured, lasted over one and a half hours (including the completion of the response booklet).

8.3 What Was Found

8.3.1 Data Reduction and the Creation of the Networks

The output of the CDM semi-structured interview is a written transcript. This was subject to a five-stage process of refinement and data reduction in order to create the DSA networks:

Stage 1 – The driver's responses to the probe questions were subject to textual analysis. This involved producing a word frequency list comprised of nouns for each participant. Nouns constitute a robust grammatical category which maps well onto the concept of information elements.

Stage 2 – When plotted on a graph, the word frequency curve approximates (visually) to a form of scree plot (e.g. Cattell, 1966). Discarding words with a frequency of less than five removed elements from the 'tail' or 'scree' of the analysis, focusing it on the most commonly and consistently occurring.

Stage 3 – The information elements (nouns) were checked in order to eliminate repetition (e.g. bike vs. bicycle). Repetitious words were combined.

Stage 4 – The noun-like elements were modelled into networks by linking them together using the same process described in Chapter 6. That is:

- One element 'has' the property of another element (e.g. [zebra crossing] <has> [traffic lights]).
- One element 'is' synonymous with another element (e.g. [traffic light] <is> [red]).
- One element 'requires' the property of another (e.g. [acceleration] <requires> [green light]).
- One element 'causes' some property in another (e.g. [traffic] <causes> [braking]).

The choice of which operator to use was based on the raw transcripts. Within these drivers literally described how one property affected another. For example, "the traffic ahead 'caused' me to brake" or "the traffic light 'is' red" etc.

Stage 5 – The networks were subject to analysis using a mixture of descriptive techniques and others drawn from social network analysis (Driskell and Mullen, 2005). Quantity of information is based on Stages #1 to #3 and a simple count of the information elements which remain in the analysis for each participant. The structure of the network in terms of its overall level of interconnectedness relies on the metric 'density'. The type of information in the network was based on the network metric 'centrality', which reveals the relative importance of different information elements.

8.3.2 Hypothesis #1: Quantity of Information

The number of key information elements for each participant was distilled from Stages 1 to 3 of the data reduction process. Figure 8.4 presents the results of comparing the between-subjects factor of group (IAM, 8 and 2 drives) and the within-subjects factor of time (pre- and post-intervention). It is apparent there are differences across

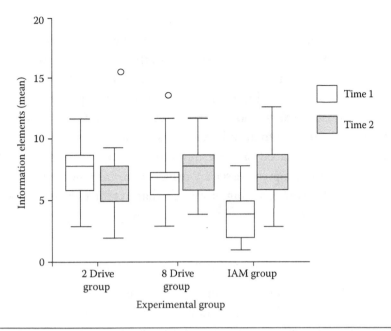

Figure 8.4 Quantity of critical information for each group and time interval.

the groups at Time 1 and next to no change in the number of key information elements in Time 2 for the control groups, but a marked increase in information quantity for the IAM group. The results of a mixed ANOVA support this observation statistically.

The mean number of information elements for the two-level time factor differed significantly and to beyond the 1% level: $F(1,20) = 6.51$; $p < 0.013$. The effect size was medium (partial eta squared = 0.08). The interaction between group and time was also statistically significant to beyond the 1% level: $F(2,72) = 9.58$; $p < 0.0001$, with the corresponding effect size being large (partial eta squared = 0.21). Figure 8.4 clearly demonstrates this interaction: the mean amount of information elements extracted from the analysis of IAM drivers increases between Times 1 and 2 by approximately 53% for a corresponding 1% increase in the 8 drive control group and a 1% decrease for the 2 drive control group. However, the IAM group's dramatic increase in informational elements only raises it up to approximately the same level as the control groups. In other words, the IAM group appears to start from a much lower base.

This would be a highly troublesome finding according to the rubric of individualistic, normative, 'situation focused' SA theories. In such cases it would be expected that IAM drivers would start at the same level as everyone else. Moreover, they would increase the quantity of information elements considerably beyond the control groups. Sliding the systems lens across the results reveals a different explanation. The effect of IAM coaching brings previously automatised information into consciousness, and any increases in the number of elements is not necessarily due to an inherently more numerous mental theory of the situation. Indeed, a more numerous situational model of the situation is something possessed by drivers who have not experienced any advanced driver coaching at all. To that extent Hypothesis #1 is supported.

8.3.3 Hypothesis #2: Structure of Information

Network diagrams showing noun-like information elements and verb-like links were created from the individual data distilled through Stages 1 to 4 of the data reduction process. The result is a set of six generic networks which characterise the three groups of drivers: a set of three networks applicable to Time 1 and a further set of three applicable to Time 2. Stage 5 of the data reduction process is then applied, the network being mathematically interrogated to extract an overall measure of its structure. Once again the metric being deployed is Density (see Chapter 6 for the mathematical formulation). Figure 8.5 presents the outcomes of applying this analysis to each of the generic networks. It can be seen how the density of the IAM group's network increases between Times 1 and 2 from 0.04 to 0.12, with no similar change evident for either control group. The implication of this finding is that following advanced driver coaching the drivers' situation models are more interconnected and developed: Hypothesis #2 is thus supported.

8.3.4 Hypothesis #3: New Information Elements

Hypothesis #3 states the SA of advanced drivers will be comprised of qualitatively different types of information. This is a result of anticipatory and feed-forward modes of car control. Some support for this is found in the following analysis.

Figure 8.6 shows the key information elements for the IAM group at Time 2 are comprised of 38% new elements (replacing 26% of

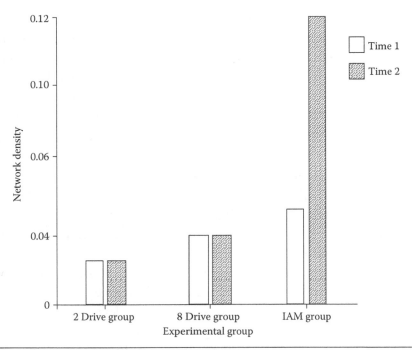

Figure 8.5 Graph illustrating changes in the configuration of information. Higher values denote greater interconnectivity between information elements.

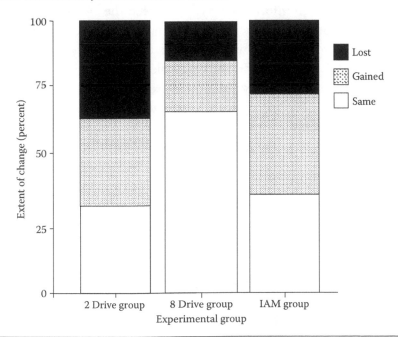

Figure 8.6 Stacked bar chart showing the percentage of information objects lost, gained and remaining the same between Times 1 and 2.

elements no longer present). The pattern is different for both control groups. For the 8 drive group around 18% of the elements present in Time 1 are no longer present in Time 2. For the 2 drive group the proportion of new elements rivals the IAM group (at 32%) but unlike the IAM group more information elements are lost between Times 1 and 2. In numerical terms, although there are clearly differences between the groups, such a simple analysis reveals little about their importance. Greater detail can be provided by applying the network metric centrality (described again in Chapter 6). The higher the Centrality score, the fewer 'hops' are needed to get from it to any other information element. Highly central elements, therefore, are rather like hubs and because of this can be regarded as important elements in the network.

When the data on informational gain/loss is re-cast in centrality terms the following emerges. For the IAM group every situational element with a centrality value one standard deviation above the mean is different at Time 2 compared to Time 1. For the 8 drive group only one such information element is different, and for the 2 drive group, only three. So, despite the comparable levels of numeric change in information type between the IAM and control groups, it appears that much of the change in information type for the control groups occurs with elements that have less importance. For the IAM group, whilst a comparable number of new information elements are gained by Time 2 (at least compared to the 2 drive control group) they are systematically more important in network terms. This finding supports Hypothesis #3. Not only is advanced driver training implicated in slightly more new information elements pre–post intervention, but unlike the control groups those new elements are more important.

8.3.5 Hypothesis #4: Old Information Elements (But Increased Importance)

Figure 8.6 shows not just the percentage change in information elements which have been gained/lost between Times 1 and 2 but also those information elements which persist throughout. Hypothesis #4 states that even though the type of information element is the same, its role in the drivers' SA will be different as a result of advanced driver coaching. Figure 8.7 shows the result of subtracting the Centrality

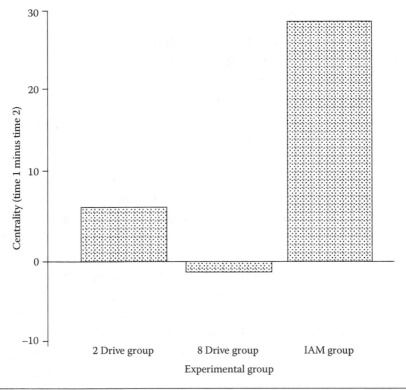

Figure 8.7 Extent of change in the criticality of situational elements common to both Times 1 and 2. Bars show the sum of the positional centrality scores after the values at Time 1 have been subtracted from Time 2.

scores for the shared information elements from each other to see, overall, how their role in the network changes. It is immediately clear that by far the greatest extent of change in centrality for shared information elements occurs within the IAM group.

Figure 8.7 shows a change in centrality of 0.28, −0.01 and 0.06 for the 2 drive, 8 drive and IAM groups respectively. For the 8 drive group there is a small positive change in centrality for shared elements, and for the 2 drive group elements that are shared become slightly less important. The findings for the IAM group, in contrast, are indicative of shared elements becoming much more central, more 'hub-like' and thus playing a more important role in defining a driver's SA. Hypothesis #4 is supported. Despite 36% of information elements remaining present in the networks between Times 1 and 2, the 'relations' between them and other elements have changed markedly.

8.3.6 Hypothesis #5: Behaviour

Hypothesis #5 is premised on the notion that advanced driving is about exerting control over the driving context via specific driving behaviours, evolving oneself towards better situations to be aware of. Reference to the wider literature provides a basis for supposing advanced drivers do this (e.g. Hoinville et al., 1972). This links to a central debate in SA: does it improve performance?

The assumption throughout this monograph is, yes, driver SA is useful. There is growing evidence that using SA principles results in improved performance. Indeed, in a recent review of the field Wickens (2008) presents a strong case for the value SA has to offer, with the number of citations and the use of the term in everyday language attesting to that. So does the varied work of Gugerty (1997/9) where SA is cited as a greater cause of traffic accidents than improper speed or technique. Interestingly, however, the relationship between SA and task performance was initially difficult to prove (Endsley, 1995). It is unquestionably the case good task performance can occur despite poor SA and vice versa. Clearly the relationship is more complex than a simple one to one SA/driving-performance mapping. Current research is providing the much needed evidence and insights (e.g. Griffin et al., 2010; Sorensen and Stanton, 2013; Sorensen and Stanton, 2016; Rafferty et al., 2013; Walker et al., 2009). These studies suggest SA, and DSA in particular, is not only a useful concept but also correlated with task success or failure. Different models of SA are also useful, but perhaps in different circumstances and for different reasons.

In this study, driver behaviour was firstly assessed on-line with the occurrence of specific behaviours being noted by the observer as the candidate/associate drove. The results of a sign test applied to the frequency data shows significant improvements in nine categories of behaviour for the IAM group. Statistically significant improvements were detected in terms of speed, speed approaching (and appreciation thereof) of bends, the application of the system of car control, steering technique, headway (following distance), use of mirrors, and gear changes (exact $p < 0.05$, two-tailed). No such improvements were detected for the control groups. In most cases, perhaps unsurprisingly, they remain unchanged between Times 1 and 2. More worryingly, however, any favourable effects of time and/or being accompanied appeared short-lived. Drivers in both control groups became significantly worse pre–post in terms of their steering technique and gear changes, and the 2 drive group became significantly worse in terms of their response to hazards (exact $p < 0.05$, two-tailed). Anecdotally, drivers in the control group tended to relax markedly to being observed/assessed and their driving style seemed to reflect this. The results are presented in Table 8.4.

The second behavioural analyses method was not based on observation of discrete behaviours and the frequency of their occurrence, but on a more subjective assessment, carried out post-drive using the IAM's existing assessment report sheet. Here 26 behaviours are ascribed a score between 1 = Excellent and 5 = Poor. The results, illustrated and summarised in Table 8.5, show how following IAM coaching there is a significant improvement in 18 of the 26 behaviour categories (exact $p < 0.05$, two-tailed). The control groups, on the other hand, either showed no change, as in the case of the 8 drive group, or else showed a statistically significant decrement.

This finding communicates that as a result of IAM coaching drivers show demonstrable improvements in their driving technique on 18 critical variables, including observation, anticipation, hazard assessment and management, speed limits, mirrors, signals and safe progress.

Table 8.4 Summary of Statistically Significant Results (and Their Direction) within Each Observed Behaviour Category

BEHAVIOUR CATEGORY	IAM	CONTROL GROUPS	
		8 DRIVES	2 DRIVES
Speed	Improvement		
Limit Points	Improvement		
Roadcraft	Improvement		
Handbrake			
Smoothness			
Signalling			
Steering	Improvement	Worse	Worse
Road Position			
Headway	Improvement		
Response to Hazard			Worse
Use of Mirrors	Improvement		
Gear Changes	Improvement	Worse	Worse

Table 8.5 Summary of Statistically Significant Results (and Their Direction) within Each Observed Behaviour Category

ASSESSMENT CATEGORY	CONTROL GROUPS		
	IAM	8 DRIVES	2 DRIVES
Acceleration			
Acceleration Sense	Improvement		
Braking	Improvement		
Clutch Control			
Gear changing			
Use of Gears			
Automatic Gears			
Steering	Improvement		
Manoeuvring			
Concentration	Improvement		
Observation	Improvement		
Anticipation	Improvement		
Hazard Assessment	Improvement		
Hazard Management	Improvement		
Road Position	Improvement		
Speed Limits	Improvement		Worse
Overtaking			
Safe Progress	Improvement		Worse
Mirrors	Improvement		
Signals	Improvement		
Road Surfaces	Improvement		Worse
Smoothness	Improvement		
Sympathy	Improvement		
Courtesy			
Aptitude			Worse
Commentary	Improvement		
Knowledge	Improvement		

The findings in favour of observation and anticipation provide further support for Hypothesis #3 and the use by advanced drivers of feedforward control. The aspects of hazard assessment and management, safe progress and speed limits all provide support for Hypothesis #5. They serve as a reflection of the type of situation advanced drivers are constructing for themselves to be aware of. It appears to be quite a different sort of situation for the 2 drive group, evidenced by the decrements in safe progress, speed limits and aptitude. Who has the better SA though? Is it the 2 drive group with their more numerous information elements, or the IAM group with their varied levels of interconnection. The answer depends on the SA lens that is slid across the frame.

8.4 Summary

In our demonstration drives with the Metropolitan Police we saw first-hand what the normative 'gold standard' of driving expressed in Chapter 4's task analysis looked like. Via the police driver's simultaneous verbal commentary we were also able to witness the key role of SA. This chapter enables us to ask whether driver SA such as this can be trained, and to burrow deeper into the concept of SA itself. What we discovered was that drivers who are subject to advanced driver coaching show an increase in the number of new information elements that comprise their SA, an increase in the overall level of interconnectivity between those elements, an increase in the criticality of new and existing elements, and an increase in favourable driving behaviours. Evidence was thus been found in favour of all five hypotheses. To this extent, it is possible to conclude that, yes, driver SA can be improved via training.

The findings also support the idea that good driver SA need not look like a mental analogue of the objective situation for it to be 'good' SA. Although the number of information elements extracted from advanced drivers increased, it did so only to the same level as the control groups. Using individual approaches to SA like Endsley's popular three-level model would be problematic. It would lead us to conclude that advanced driver coaching makes SA worse, despite driving performance becoming better. It would also steer the argument towards the idea SA might not actually be useful, that it was a poor (or even negative!) predictor of task outcomes. This is a clear sign a contingent approach to SA is needed. SA approaches need to be matched to key aspects of SA problems. In this case it seems that a normative individual model of SA is able to address some research questions better than others, such as those in Chapter 5.

A systems view of SA seems to provide much more adequate answers to the distinctly non-linear questions being asked in this chapter. An increase in the interconnectivity of knowledge is more important and relevant than an increase (or decrease) in the number of discreet information elements. By focusing on the structural determinates of information networks, the situational models held by the individual and, indeed, system, can be assessed in ways which are to some extent situation independent. On first impressions this might seem paradoxical. Yet, it is certainly the case here that it is not what SA 'is' as such, but how it is 'constructed'. Features of that 'construction' are what the current method has diagnosed and it is these which seem to link to improved performance, not merely the simple quantity of information.

This network-based approach to the measurement of driver SA is compatible with, and responsive to, long standing knowledge about expert decision making, specifically, that experts chunk information and look for patterns of interrelationships: SA is more than the sheer quantity or even type of information possessed by an individual. The network-based approach is also flexible; it is not tied to any particular knowledge elicitation paradigm, nor is it focused just on individual cognition. It can just as easily represent systems-level SA in which different individuals own and/or share knowledge, as it can distributed and joint cognitive phenomena, in which situational elements are held, represented or otherwise transformed by non-human elements of a system. It also begs the question of how future vehicles could help 'normal' drivers create the SA, and better situations to be aware of, that advanced drivers seem able to achieve. Could some forms of vehicle technology be simply removing basic and operational driving tasks when, in fact, the real benefits are to be accrued from 'assisting' the driver with tactical and strategic tasks? Could some trends in vehicle design be hindering the acquisition of SA? Could a technology be envisaged which bestows drivers with the SA achieved from advanced driving techniques without the need for particular advanced driving skills? This represents an intriguing challenge which, in some important respects, sounds quite different to where we might be currently heading: a form of automation which gives every driver an IAM observer or, better still, a police driving instructor.

9

Driver SA and the Future City

9.1 Introduction

9.1.1 Journey into Mega-City One

"Each year, more than 100,000 people descend on San Diego for Comic-Con International. The largest annual comic and pop culture convention in the world. San Diego Comic-Con offers fans a chance to immerse themselves in the world of their favourite superheroes, with panels, previews and promotions featuring renowned actors and comic book professionals. This year, The Conversation [an independent source of academic news and views delivered direct to the public] *has given one academic the chance to do the same."* The academic in question happens to be one of the authors of this text.

It is not often one gets the opportunity to raise oneself above the day-to-day realities of grant applications, university admin and teaching to instead journey into the imagined world of comic-book cities. Neither is it common to embed human factors and ergonomics within them. Having said that, these often-grotesque urban dystopias seem to owe a great deal to a distinct lack of human-centredness. Indeed, on closer inspection, many of the popular 'future city' tropes serve as entertaining examples of how current modes of thought can give rise to paradoxical ergonomic outcomes.

In the present case our attention turns to driver situation awareness (SA) not at an individual, team or even systems level as normally encountered, but to SA at a city scale. This is a considerable scaling up of the ergonomics system's agenda (e.g. Dul et al., 2012) but why not? Vehicle-to-vehicle and vehicle-to-infrastructure communications, embedded computing, artificial intelligence, smart cities and the internet of things actively require this level of analysis (Walker et al., 2017). Indeed, the key to their success lies increasingly in the human dimension which we ignore at our peril. On these terms one comic-book city presents itself as the exemplar, a city which, despite being imaginary, came top in the *Architects' Journal*'s list of Top 10 comic-book cities: Mega-City One, from Judge Dredd fame. In this penultimate chapter we invite the reader to join us on an ergonomic flight of fancy into the future, one backed up by a fascinating study into the effects of driver SA at a city scale, and the paradoxical disconnects between individual, team and systems SA.

The popular science article which appeared in *The Conversation* (accessible here: https://theconversation.com/from-self-driving-cars-to-zoomtubes-an-expert-imagines-the-evolution-of-transport-in-mega-city-one-81237) describes how something has been found underneath Sector 301 of Mega-City One. Judge Dredd is on his way to the scene:

Coming in from above, shiny Zoomtubes can be seen weaving their way through the monolithic habitation blocks and unbroken urban blight. They pulsate with computer-controlled convoys of fast moving automated vehicles speeding along inside a vacuum. 800 million people live in Mega-City One. It's crowded. Convulsing. Choking. Breaking under its own weight. The civilian population are mostly illiterate because artificial intelligence removed the need for most types of work, but they are restless, always on the move and often in trouble. This is why street judges like Dredd exist. To dispatch instant justice and forcibly restore order; they are the law.

Mega-City One has a secret. It is built over the top of abandoned and ruined 'under cities' from before the 2070AD nuclear war. Dredd is descending into this dark undercroft now. Spotlights have been set up around a crime scene but this is not what attracts Dredd's attention as his Lawmaster motorbike parks itself. No. Behind an open roller-shutter door in the side of an ancient concrete building from 1970AD is what appears to be a brand-new petrol burning vehicle. These were mass-produced in the twentieth century. This, then, is an unbelievably rare and expensive antique. Why it is here is unclear. It sits in what appears to be a laboratory of some kind, connected to ancient silicon-based computers. A transport professor from the City Central De-Education Establishment is sitting inside the vehicle looking around in bemused wonderment. She says, "They were trying to steal this. It's a completely intact driving simulator laboratory from the twenty-first century." Dredd thought for a moment: "What is driving?"

The professor chuckled. "150 years ago people would sit here and turn this large round thing in front of them to send the vehicle left or right, and press these pedals here to start and stop." Brushing some cobwebs away from the top of the instrument panel the professor went on: "Sounds dangerous doesn't it. And in some ways it was. It was quite amazing, though, how something so primitive could be used by so many people." She reached into the passenger seat and picked up a dusty thick folder containing hundreds of paper sheets. "People used to think driving was a simple activity but this is a task analysis. It shows people – not artificial intelligence – had to perform over 1600 individual tasks, at the correct time, in the correct sequence, in order to avoid crashing this thing. Amazing really. In fact, people had quite a lot of trouble adapting to automatic vehicles." Dredd looked at her incredulously. "I know!" She shifted herself out of the driver's seat and walked over to the other side of the room. Dredd followed, intrigued.

Part of the roof had collapsed and water was leaking in, dripping on piles of old paper books and broken coffee mugs with the crest of a once famous university printed on them. The professor crouched down and peeled away a thin sheaf of water-damaged paper from the pile. "This will take years to go through, but look at this. Look at how people in the twenty-first century were thinking about vehicle automation. They categorised it into six levels, from zero automation – a bit like that petrol burning vehicle over there – where the driver does everything, even steering, right through to full automation like we have now. What's interesting are the levels in between. Their AI wasn't sophisticated enough for full automation in all conditions so the vehicle did some of the control but the human driver had to do the rest, and judging by all these other ancient texts lying here that caused no end of trouble."

9.1.2 Intelligent Transport Systems

Mega-City One: an implausible dystopian vision? Perhaps not. As of the early twenty-first century approximately 90% of the UK population now lives in an urban area; this is a trend mirrored across the world. Urbanism drives increasing traffic movements within these areas. These movements interact with dense street networks and a strong planning incentive to maximise their capacity and reduce vehicle emissions (Hesse and Rodrigue, 2004).

Intelligent Transport Systems (ITS), in the form of route guidance and advanced traveller assistance systems, are seen as a key enabler for this.

There is good evidence for the positive benefits ITS of this sort can have (e.g. Asvin, 2008; Giannopoulos, 1996; Dutton, 2011 etc.) but as the technology continues along the s-curve towards full market saturation there are some fundamental questions which still need to be explored. For example, are some urban road network topologies more energy efficient when paired with ITS technology than others? If so, to what extent might this influence a wider ITS strategy? Do all drivers have to have complete knowledge of traffic conditions, and therefore perfect SA? How realistic is this assumption anyway? Is it safe to assume that having invested in ITS drivers will adhere to route guidance information in all cases? Research (e.g. Lyons et al., 2008; Bonsall and Palmer, 1999; Bonsall, 1992; Chorus et al., 2006; Karl and Bechervaise, 2003 etc.) shows that between 30% and 50% of drivers do not: what happens then? Do these drivers end up with partial SA? Clearly, the success of some prominent types of ITS technology are heavily contingent on SA. This study tries to provide a thought-provoking exploration of the issue before we descend into an anti-human factors Mega-City One abyss from which we cannot escape.

9.1.3 Street Patterns and Network Types

The language and metaphors of systems thinking are used in transport network analysis just as they are used to map out networks of information and SA. Systems thinking is, after all, a universal language. In transportation research network representations like this are used to reduce a complex transport network into a set of fundamental elements: nodes that represent junctions and links that represent roads (Lowe, 1975). A two-dimensional set of systematically organised points and lines like this are referred to as a planar graph. These are the basis upon which various forms of spatial analysis quite often proceed (e.g. Bowen, 2012; O'Kelly, 1998). Urban street networks tend to evolve as a product of the area's rate of growth, period of formation, location, topography, climate, culture and so on (Thomson, 1977). Planar graphs of street patterns reveal their individuality but also their common patterns or archetypes. Mega-City One's topology is rather unclear; likely candidates might be 'megaform' or 'polycentric net', but there are many others besides as Table 9.1 shows.

A long-standing goal in transportation research has been to define certain road network 'typologies' (e.g. Reggiani et al., 1995), a task made more difficult by the inconsistent use of terms such as those

Table 9.1   Descriptive Terms Applied to Urban Street Patterns

AUTHOR	DESCRIPTIVE TERMS
Unwin (1920)	Irregular
	Regular
	Rectilinear
	Circular
	Diagonal
	Radiating lines
Moholy-Nagy (1969)	Geomorphic
	Concentric
	Orthogonal-connective
	Orthogonal-modular
	Clustered
Lynch (1981)	Star (radial)
	Satellite cities
	Linear city
	Rectangular grid cities
	Baroque axial network
	The lacework
	Other grid (parallel, triangular, hexagonal)
	The 'inward' city (medieval, Islamic)
	The nested city
	Current imaginings (megaform, bubble floating, underground, etc.)
Satoh (1998)	Warped grid
	Radial
	Horseback
	Whirlpool
	Unique structures
Frey (1999)	The core city
	The star city
	The satellite city
	The galaxy of settlements
	The linear city
	The polycentric net

Source: Marshall, S., *Streets and Patterns*, Oxon and New York, Spon Press, 2005.

shown in Table 9.1. Despite this, it is possible to discern a much smaller recurring subset of network patterns. Brindle (1996) argues for as few as two 'fundamental' urban street layouts, the grid and the tributary. Table 9.1, however, aligns with the work of Marshall (2005) in which four 'archetypal' street patterns can be identified, the 'linear', 'tributary', 'radial' and 'grid' patterns, shown graphically in Figure 9.1.

9.1.4 Network Metrics

Network diagrams provide a visual representation that, in simple cases, makes it very easy to discern one archetype from another. In more complex real-world examples the visual complexity makes this task difficult to perform reliably and objectively. In these cases it is possible to turn to a number of formal metrics drawn from graph theory. These enable the connectivity of street patterns to be calculated in a similar way to those which are applied to information networks. In this case three metrics are used: the 'Beta Index', the 'Gamma Index' and 'Network Depth'.

The Beta Index is a simple equation, somewhat similar to network centrality (see Chapter 7). It is used to determine the relationship between the total number of links and the total number of nodes in a network. It is calculated using the following equation:

$$\beta = e / v$$

where,

e = number of links
v = number of nodes

$$0.5 < \beta < (v - 1) / 2$$

The Beta Index provides a measure of linkage intensity or "the number of linkages per node" (Lowe, 1975). Beta values generally lie between 0.5 and 3. Networks of values > 1 consist of some nodes having more

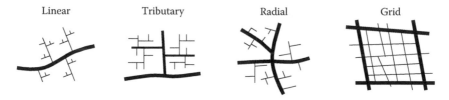

Figure 9.1   Marshall's (2005) urban street pattern archetypes.

than one route between them, and the network would be considered well connected. Beta values < 1 indicate the network is not as well connected and that there are only singular routes between nodes.

The 'Gamma Index' helps to identify "the ratio between the actual and the maximum possible number of links" in the network (Lowe, 1975). This essentially determines whether or not every node is connected by a link, and is conceptually similar to network density (see again Chapter 7). Gamma is derived from the following equation:

$$\gamma = e \,/\, \big(3(v - 2)\big)$$

where

  e = number of links
  v = number of nodes

$$0 < \gamma < 1$$

In this case, if γ = 0.5 it means only 50% of the maximum possible number of links in a network are present, and that all of the nodes are not fully connected.

The 'depth' of a network is a relatively simple concept. In essence, it accounts for the relative distance between the most minor route and the most major route. It establishes the idea of a hierarchy and the different interconnections which exist between levels. For example, minor roads providing access to houses are 'deep' whilst major routes are 'shallow'. An example of this is shown in Figure 9.2 where we see that the thinner the link, the deeper the road. This particular network, taken from Marshall (2005), shows the network has a depth of five.

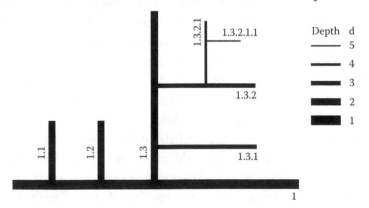

Figure 9.2   An example of what is meant by the 'depth' of a network. The major route is shown as the thickest line with the minor routes represented as thinner lines.

Network depth is important if the goal is to impartially compare different street patterns. If one network was to have a depth of four and another a depth of two, the 'deeper' network would have more connectivity. This would allow the road user to travel between points in a shorter period of time because there are more route choice options for them to exploit. Different street patterns (with different Beta and Gamma coefficients) can, however, be legitimately compared if the network depth is kept constant.

9.1.5 *The Problem*

Abstracting street patterns to the level of planar networks puts them on the same level as communications and other network types which, as we have already seen, puts them in contact with a rich literature in the fields of graph theory (Harary, 1994), sociometry (Leavitt, 1951; Monge and Contractor, 2003) and complexity (Watts and Strogatz, 1998). It also aligns them in interesting ways with the systems-level views of SA discussed in the previous chapters: they too use networks and network metrics. We know from the wider network literature that network type is a strong contingency factor in how they perform (Leavitt, 1951; Pugh et al., 1968; Watts and Strogatz, 1998 etc.) and the same is true of street networks. The question arises once again as to whether 'more' SA is necessarily better. By the same token, more technology and in this case more route guidance may not be better either (Nijkamp et al., 1997). We know this from the wider literature. Of course, roads are not communications networks in the way they have been previously studied. So, in order to explore the hypothesis that street pattern is an important contingency factor in the benefits to be accrued from ITS-induced improvements in SA we instead apply an agent modelling approach. The approach is called *traffic microsimulation*. Microsimulation enables us to create a set of virtual street layouts based on real towns and cities, populate them with virtual traffic having differing levels of ITS-induced driver SA, and observe the outcomes in terms of journey length, duration, cost and carbon emissions under different conditions. This is a highly exploratory study aimed at pushing systems ergonomic ideas further than they have been before in order to see what emerges. Is it the dystopian vision evoked by comic-book cities like Mega-City One or something else?

Looking Judge Dredd in the eye the professor continued: "I wonder what the inhabitants of this ancient city would have thought of Mega-City One." "They would have learnt a lot from our advanced technology," Dredd replied. Turning away a little wistfully the professor thought out loud: "I'm not so sure. Mega-City One is like a giant machine. The technology rules. It is a logical extension of the ways we used to think about cities and transport at the time this old building was built in 1970AD." She waved her hand vaguely in the direction of the decaying concrete structure they were standing in. "Maybe we could have taken a different direction. A more human-centred direction in which people and technology are jointly optimised and mutually reinforcing."

9.2 What Was Done

9.2.1 Design

This study used traffic microsimulation to test the interaction between different 'amounts' of driver SA and the outcomes achieved within different street patterns. There are four dependent variables: journey duration, journey length, journey cost and carbon output. These dependent variables are contingent on two independent variables: amount of driver SA and street pattern type. The street patterns were based on real urban locations, with network depth held constant in order to control for non-systematic biases in connectivity. Network demand was also based on real-life traffic count data from the relevant sites.

9.2.2 Real-Life Urban Networks

As determined in the previous section, there are four common forms of urban transport network layout: linear, radial, grid and tributary. It was necessary to develop microsimulation models which closely reflect the layouts of these urban networks and while fun to consider comic-book cities of the future, for experimental purposes it is more useful to relate the layouts to real-life towns and cities. The Beta and Gamma coefficients were used to calculate the connectivity of the network archetypes shown in Figure 9.1 in order to select real-life road networks exhibiting the same properties. This approach allows networks to be categorised by a visual and a statistical approach. The depth of each model was set to three in order to control for the effects of network magnitude.

The Linear network required locating a small town featuring just one main road with the town located along its length. A settlement identified as meeting these criteria is probably the polar opposite of Mega-City One. It is the altogether more charming Aviemore, located in the Scottish Highlands. The Radial network is one that has several roads intersecting or converging at a centre, analogous to the spokes of a bicycle wheel. These features can often be found in a larger town which has been formed at a crossroads. These criteria are met by Dalkeith, a town located just to the south of Edinburgh. The Grid network, as its name suggests, is a network with straight roads intersecting other straight roads at right angles to form a collection of squares or blocks. The characteristics of this network type are found in the centre of Glasgow. Finally, the Tributary network is analogous to tributary rivers, with the smaller rivers feeding the bigger rivers. In road networks, it is the small roads that connect to the larger roads with 'Network Depth 1' only connecting to 'Depth 2' and 'Depth 2' only connecting to 'Depth 3' and so on. The result of this is that only the shallower roads are busy and the deeper roads are not used as shortcuts. An area meeting the description of a Tributary network is Livingston. This is also not Mega-City One but it is a so-called 'new town' in central Scotland, planned and designed in the 1960s. Planar graphs of these real-life street networks are shown in Figure 9.3.

Beta and Gamma values are calculated for each of the real-life towns to check the extent to which they conform to the linear, radial, grid and tributary archetypes, as shown in Table 9.2. The real-life networks are considerably larger than the archetypes yet it can be noted how the network metrics align, showing how the underlying structures are equivalent, independent of network size. The Beta values for the 'Linear' and 'Aviemore' networks, for example, are exactly equal as are the Beta values for the 'Tributary' and 'Livingston' networks. Even where differences do exist (and they are relatively modest) the rank order of the network types is still preserved.

9.2.3 Development of Network Models

Planar graph representations of Aviemore, Dalkeith, Glasgow and Livingston were extracted from ArcGIS, a mapping and spatial analysis software, and imported into S-Paramics, the traffic microsimulation software, in order to create a basic model of each. An attempt was

Aviemore (Linear)   Dalkeith (Radial)

Glasgow (Grid)   Livingston (Tributary)

Figure 9.3   Traffic microsimulation models of Aviemore (linear), Dalkeith (radial), Glasgow (grid) and Livingston (tributary).

made to ensure the modelled network was as true to life as possible. Nevertheless, some simplifications were required to isolate the effects of street patterns and route guidance from specific and localised variations in the traffic situation. As such, junctions had simple priorities applied with the bigger roads having priority over the smaller roads. Traffic lights were avoided all together. There were also no buses or bus routes applied as not all networks had buses arriving at similar times and some networks had bus lanes whereas others did not. A number of steps were taken to calibrate the models to the real-world context despite these simplifications being applied. A number of parameters in the model were adjusted to ensure this. Firstly, the Signpost Distance, which refers to the distance ahead of a hazard the modelled road users become aware of it, was standardised to 80 m.

Table 9.2   The Values Obtained for Beta and Gamma for the Standard Layouts Shown in Figure 9.3

	NETWORK	NODES (V)	LINKS (E)	BETA (β)	GAMMA (γ)
Archetype	Linear	19	35	1.84	69%
Real Life	Aviemore	140	257	1.84	62%
Archetype	Radial	32	65	2.03	72%
Real Life	Dalkeith	140	268	1.91	65%
Archetype	Grid	33	73	2.21	78%
Real Life	Glasgow	140	278	1.99	67%
Archetype	Tributary	9	17	1.89	81%
Real Life	Livingston	140	265	1.89	64%

This allowed the traffic to use both lanes of the entrances and exits of roundabouts (anomalous behaviour would arise if not). Visibility was also varied. This relates to how far back from the stop line simulated vehicles begin assessing their gap distance. This ensured vehicles approaching roundabouts continued onto it in situations where nothing was coming and stopped in cases where something was approaching. The distance at which vehicles followed this rule was found to be 20 m. These rules allowed more life-like driver behaviour to be represented, and for a good approximation of the traffic situation actually experienced in these locations to emerge.

9.2.4 Driver SA

The principle SA-based manipulation was the amount of knowledge drivers had of wider network conditions. This was based on the amount of ITS-mediated route guidance provided by the vehicle. Three SA 'amounts' were implemented:

- SA Amount 1 represented a driver who had 100% knowledge of traffic conditions on the network via ITS and followed the route guidance information they were given 100% of the time. This driver group is referred to as having 100% SA.
- SA Amount 2 represented a driver who had 50% knowledge of traffic conditions on the network via ITS. This driver group is referred to as having 50% SA.

- SA Amount 3 represented drivers who had no ITS or route guidance. The drivers had 0% knowledge of traffic conditions on the network beyond what they could see ahead of them. This driver group is referred to as having 0% SA: not literally zero (they are aware of their immediate situation) but zero in terms of wider network performance and opportunities to use it differently.

The three amounts of virtual driver SA were implemented in the microsimulation model by manipulating a variable called feedback. In this case feedback is information supplied to the road user, via ITS, about the current network conditions. Specifically, it lets them know of journey times on all routes so they can decide an optimum route to choose. Feedback is calculated using two aspects: the feedback interval and the feedback factor.

The feedback interval refers to how often the information is updated. An interval of two minutes was used in the study. This is a common value in similar models and avoids the modelled drivers making unrealistically rapid and unstable route choice decisions.

The feedback factor is about what percentage of delay information is taken from the previous feedback interval. The value was set as 0.5, a standard value in traffic models of this kind.

A perturbation level of 5% was applied to all three driver SA types. This helps to account for variability in travel costs, or a driver's perception of these travel costs. As perturbation increases, the virtual road users' concentration tends to focus more on reducing journey cost. However, applying a small percentage means road users will continue to focus on reducing journey length and journey duration as the key variants but will also look to reduce their cost simultaneously.

The same mix of traffic was applied to all the networks. 50% of the road users in all the network types were 'SA Amount 3' (0% SA) with the remaining 50% being split in two: 25% of road users were 'SA Amount 2' (50% SA) and 25% of road users were 'SA Amount 1' (100% SA). The traffic was modelled as medium-sized petrol-powered cars (rather than Judge Dredd-style MoPads or Lawmasters, sadly).

9.2.5 Network Demand

The demand profile controls the number of vehicles in the network, the origin and destination of the vehicles, the vehicle type percentages, and the release rate of vehicles into the network. The number of vehicles released from the origin to the destination was varied from model to model based on actual traffic count data. The goal was to bring each network to its peak PM traffic flow and hold it there for the duration of the study. To do this, peak PM values from the nearest traffic counter site were used and multiplied by 24.

The release rate of the vehicles over the 24-hour period was constant. The modelled vehicles were not released at once and a steady flow was maintained. Each model was then subject to 30 'batch runs' between the network model hours of 1600 and 2000, which was the length of the PM peak flow period. The models had, of course, been established in this peak flow state for many hours previously hence any transient effects of the model being initially loaded with traffic were avoided. These 30 runs allowed a significant amount of data to be obtained on the outcome variables of journey duration, length, cost and carbon output and their contingency on network type and driver SA.

9.3 What Was Found

The output from running the S-Paramics microsimulations is a set of raw data for each individual vehicle as it progressed through the network. The software also calculates the position of each individual vehicle every half-second and records its co-ordinates against time. This data was passed through an external programme known as 'AIRE' which calculated the resulting pollution output of each individual vehicle based on a specific year, which was chosen to be 2012. An average value for carbon output was obtained for each vehicle type, along with data on journey duration, length and cost. A summary of the uncorrected average and standard deviation values obtained from the simulations is provided in Table 9.3. Table 9.4 shows the same results corrected for network size/distance, collapsing some of the key variables into average speed, cost per km, carbon output in g/km and total emissions.

Table 9.3   Summary of the Uncorrected Journey Length, Duration, Cost and Carbon Output Data Obtained from the Modelled Networks

	SA (%)	JOURNEY LENGTH (metres)		JOURNEY DURATION (seconds)		JOURNEY COST (pence)		CARBON OUTPUT (grams)	
		Mean	SD	Mean	SD	Mean	SD	Mean	SD
Linear	0%	2152.7	30.7	183.3	2.4	224.6	1.6	130.0	1.8
	50%	2155.3	51.6	182.7	4.0	225.2	2.5	128.5	2.9
	100%	2165.8	44.2	182.8	3.4	226.0	2.6	128.1	2.7
	Mean	2157.9		182.9		225.3		128.9	
Radial	0%	2286.4	10.5	194.6	0.8	198.8	0.8	137.7	0.8
	50%	2281.2	16.9	193.7	1.3	198.2	1.1	137.2	1.3
	100%	2285.8	17.5	193.7	1.3	198.9	1.1	137.2	1.1
	Mean	2284.5		194.0		198.6		137.4	
Grid	0%	1304.6	6.3	147.5	0.8	124.2	0.6	129.0	0.9
	50%	1316.3	9.5	142.5	1.3	124.2	0.9	122.1	1.3
	100%	1326.9	12.2	137.5	1.2	124.0	1.0	115.0	1.0
	Mean	1315.9		142.5		124.1		122.0	
Tributary	0%	3662.4	16.2	309.8	1.2	321.5	1.6	214.5	1.0
	50%	3654.8	28.7	309.0	2.4	320.6	2.9	213.8	1.6
	100%	3663.6	26.0	309.4	2.2	321.4	2.9	214.4	1.9
	Mean	3660.3		309.4		321.2		214.2	

The values shown in Table 9.3 are absolute, in that no correction is made for the differing sizes of the networks. The values in Table 9.4 are corrected and provide a number of key insights. Ignoring driver SA for a moment, it can be noted how the linear and radial networks are very similar, both being able to support average speeds in the region of 42 km/h and costs per km of approximately 27 pence, with vehicles on the network each emitting approximately 60 g/km of carbon. The tributary is the same aside from cost, which is the lowest of all the road networks at 15.99 pence/km. The slowest (33.28 km/h), most expensive (70.53 pence/km) and carbon-intensive (92.77 g/km) network is the 'Grid'. However, it is within this network that the most situationally aware drivers performed the best. In this situation 'perfect SA' is raising average speeds by 2.9 km/h, reducing costs by a not-insignificant 10.47 pence/km and, most importantly, reducing carbon

Table 9.4 Summary of Corrected Speed, Cost and Carbon Output Data

	VEHICLES (Total *N*)	SA (%)	SPEED (km/h)	COST (pence/km)	CARBON (g/km)	(total kg)	
Linear	3000	0%	42.28	28.05	60.39	390.00	
		50%	42.47	27.66	59.62	385.50	
		100%	42.65	27.31	59.15	384.32	
		Mean	42.47	27.67	59.72	**1159.82**	Total
Radial	4752	0%	42.30	26.34	60.23	654.40	
		50%	42.40	26.34	60.14	651.93	
		100%	42.48	26.26	60.02	651.95	
		Mean	42.39	26.32	60.13	**1958.28**	Total
Grid	6000	0%	31.84	75.79	98.88	773.99	
		50%	33.25	70.47	92.76	732.60	
		100%	34.74	65.32	86.67	690.02	
		Mean	33.28	70.53	92.77	**2196.61**	Total
Tributary	5995	0%	42.56	15.99	58.57	1285.97	
		50%	42.58	16.00	58.50	1281.77	
		100%	42.63	15.97	58.52	1285.29	
		Mean	42.59	15.99	58.53	**3853.03**	Total

emissions by 12.21 g/km. The carbon value is of course determined by the physical size of the networks and the journey lengths therein, so in these examples the larger Tributary network (i.e. Livingston) has the highest total emissions (3.8 tonnes of carbon per modelled PM peak); however, the network with the shortest average journey lengths (i.e. Glasgow / Grid) has the second-worst carbon outputs (2.2 tonnes).

There are smaller differential effects present in the other network types that are also important. Although smaller at the level of individual vehicles, when multiplied by the number of vehicles in the networks (several thousand) and the number of times PM peak hour conditions occur (every weekday evening) these differences begin to magnify significantly. For example, in the Tributary network, a per-vehicle difference in carbon emissions of only 0.7 grams as a result of driver SA still multiplies to an additional daily PM peak carbon output of approximately 4.2 kg, or approximately 1 tonne per year. Multiplied again by the number of settlements with tributary street patterns, these initially marginal differences start to accumulate rapidly. With this in mind, the following sections shift the focus from absolute values to relative values in order to discern the direction of these various effects, and what they might mean for a driver SA strategy under the auspices of a future smart city of some kind.

9.3.1 Driver SA versus Journey Duration

In order to provide a visually tractable representation of how the data is behaving, and show the relationships for each network on the same graph, it is necessary to convert absolute values to relative values. These were determined by finding the difference between the actual value and the minimum value and dividing this by the difference. This, therefore, shows the direction of the relationships between road network type and how different amounts of driver SA within it perform.

Figure 9.4 shows the relationship between different levels of ITS-induced driver SA and the relative journey duration experienced by each driver for each of the four networks. The graph shows that drivers with 100% SA, operating within Linear and Tributary road networks, are worse off than drivers with only 50% SA, albeit not as bad as the population of drivers with no SA. The results show that the same outcomes on journey duration are achieved at both 100%

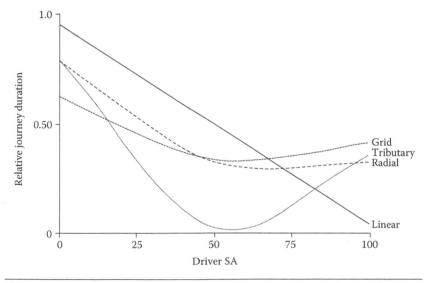

Figure 9.4   Interactions between street pattern, driver knowledge of network conditions (i.e. SA) and relative journey duration.

driver SA and approximately 20%. The difference, of course, is that the latter is likely to be considerably easier and less costly than the former to implement. The trend line obtained for the Radial network only slightly differs from the Linear and Tributary networks. Journey duration increases between 50% and 100% driver SA, levelling off thereafter. This shows there is no further benefit of improved individual driver SA in this context. The relationship for the Linear, Radial and Tributary networks seem to show there is an optimum level of driver SA. For the optimisation of journey durations 100% driver SA is not required. This principle does not hold for the Grid network. Here the trend line shows increasing benefits as driver SA increases, reaching a maximum benefit at 100% SA. The broader principle to be extracted here is that 50% SA of the traffic conditions on the network are optimum for networks characterised by one or two critical routes between the majority of the origins and destinations. 100% SA reduces journey durations in networks where there are a large range of routes available to the road user. In this study, drivers with 100% SA extract maximum journey time benefits in Grid networks.

9.3.2 Driver SA versus Journey Length

Figure 9.5 shows how the journey length taken by road users relates to their SA, but is also contingent on street pattern type. In the linear network, the journey length increases at a slow rate between 0% and 50% driver SA before it increases at a much greater rate between 50% and 100% SA. This relationship is very different from the Radial and Tributary networks, which both have a long journey length for road users with 0% SA but fall markedly between 0% and 50% SA, before rising again very steeply between 50% and 100%. As with the results for journey duration, there is an optimum level of driver SA for Radial and Tributary networks (around 50%), with the same network performance achieved at 0% SA as at 100%. In other words, the effort and expense of bestowing drivers with 100% SA via ITS would not be worth it in this instance. Strategically, this sort of insight is worth knowing. The Grid network differs though. The diagonal line traced through the chart shows the greater the amount of driver SA that is observed, the greater the journey length. This is because drivers use their enhanced SA to exploit the more numerous opportunities to

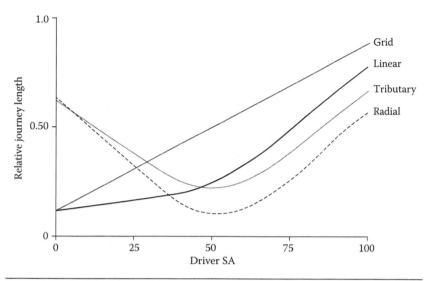

Figure 9.5   Interactions between street pattern, driver knowledge of network conditions (i.e. SA) and relative journey length.

divert. Taken together, Figures 9.4 and 9.5 suggest that shorter journey durations are achieved with longer journey lengths. The broader principle to be extracted is that vehicles in networks with more than one route between the origin and destination travel a greater distance but in a shorter time.

9.3.3 Driver SA versus Journey Cost

Figure 9.6 shows the Linear network has a more or less direct relationship between driver SA and journey cost (i.e. a straight line). It shows that as driver SA increases from 0% to 100%, the journey cost also increases. Interestingly, then, in this network type 0% driver SA is the optimum value for cost to be optimised. This is not the case for Radial and Tributary networks. Both of these undergo a reduction in cost between 0% and 50% SA, with an increase in cost then occurring between 50% and 100% SA. Once again, the optimum level of driver SA in a Radial or Tributary network is 50%, with further increases not only having a negative effect on journey cost but 100% driver SA yielding the same outcome as 0%. The Grid network again performs differently to the other three network types. The relationship is linear (i.e. a straight line) between 0% and 50% driver SA, before tailing off slightly as 100% SA is reached. What this means is that 100% SA is required in Grid networks for meaningful

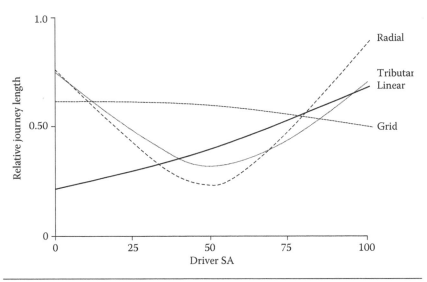

Figure 9.6 Interactions between street pattern, driver knowledge of network conditions (i.e. SA) and relative journey cost.

journey cost savings to emerge, with 0% SA required for linear networks and 50% SA for Radial and Tributary networks.

9.3.4 SA and Carbon Emissions

The crux of the analysis was to see what effect these contingent values of journey length, duration and cost ultimately have on carbon emissions. Figure 9.7 presents the results of this analysis. The Linear, Radial and Tributary networks follow a similar trend, with carbon emissions decreasing rapidly between 0% and 50% driver SA, and varying levels of diminishing further benefits as driver SA approaches 100%. The Linear and Radial networks level off beyond 50% SA, suggesting little (if any) further benefits of increasing it. The results suggest that the carbon emissions from the Tributary network worsen with increases beyond 50% driver SA. The Grid network is once again quite distinct. It contains a directly proportional relationship between driver SA and carbon emissions, with the maximum value occurring at 0% SA and the minimum value at 100% SA. If reducing carbon emissions is one of the goals of ITS then it has the biggest role to play in Grid networks. Apart from Tributary networks, there are benefits to be extracted sometimes well before 100%.

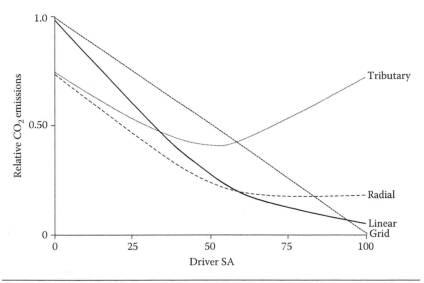

Figure 9.7 Interactions between street pattern, driver knowledge of network conditions (i.e. SA) and relative CO_2 emissions.

9.3.5 Optimisation of Driver SA

Based on the data and relationships obtained from the previous sections it is possible to create Table 9.5. This shows the potential network performance trade-offs involved in minimising carbon emissions using ITS-induced improvements in driver SA. A mean level of driver SA is given, representing a simple value by which the best compromise of journey duration, length, cost and carbon variables is achieved.

Table 9.5 shows that, in theory, urban street networks can be designed to be sustainable with the smallest pollution outputs, designed to reduce the journey duration of road users travelling through them, or designed to reduce traveller costs. In practice, however, street networks have evolved and cannot be changed on the scale necessary to optimise these factors. This is where ITS comes in. ITS interacts with network types to modify their inherent performance. The interaction is not a simple one. It is contingent on the level of information provided to, and accepted by, drivers, combined with the topology of the network itself. Where some networks require varying values of driver SA to optimise various aspects of the network, some run at the optimum level for the majority of characteristics under one

Table 9.5 Optimum Levels of Driver SA for Journey Duration, Length, Cost and Carbon Emissions for Linear, Tributary, Radial and Grid Street Patterns

NETWORK	DURATION	LENGTH	TRAVELLER COSTS	CO_2	MEAN
Linear	50%	0%	0%	100%	38%
Radial	50%	50%	50%	100%	63%
Grid	100%	0%	100%	100%	75%
Tributary	50%	50%	50%	50%	50%

level of driver SA. This can be seen in Table 9.5 with the Tributary and Radial networks, which both run at their most efficient levels for all four characteristics with a 50% driver SA amount. The table also shows that for a Grid network, 100% driver SA is the optimum value. Linear networks, however, do not reach their optimum level of efficiency for any one amount of driver SA. Table 9.5 becomes a useful tool when thinking and reasoning about future smart cities. Stated simply, while the maximisation of driver SA at the individual or team levels might accrue benefits for the individuals or teams concerned, the collective benefits for the system as a whole might be quite different. Some levels of driver SA are more optimal than others, and it depends on the street pattern those drivers are operating within.

The results discussed above convey the idea that for each output characteristic (i.e. journey length, cost, duration and carbon emissions) different drivers will perform differently depending on the level of ITS-induced SA. But what if a particular context – like a future city – does not conform to the archetypes presented? In this case it is possible to increase the generalisability of the results with recourse back to the connectivity coefficients discussed earlier. These can be applied to any transport network, of any size or type, real or imagined, in order to reveal its underlying level of connectivity. A fundamental relationship emerges: as driver SA increases from 0% to 100% in a road network with Beta (β) ≥ 1.9, journey costs, journey duration and carbon output tend to improve (or at least do not worsen). In networks with $\beta < 1.9$ there is no added benefit of supporting anything more than 50% driver SA. Figure 9.8 illustrates the relationship between β and driver SA within the range of data observed. Figure 9.8 represents a simple diagnostic for answering the question, 'What are the benefits of increasing drivers' SA of the wider network conditions, via ITS, in a particular operating environment?' Those benefits are clearly complex and non-linear.

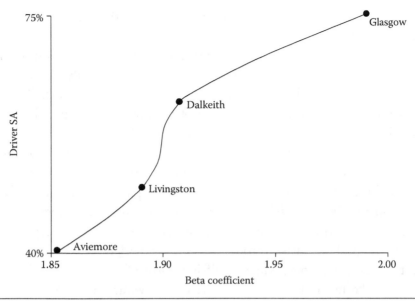

Figure 9.8   Relationship between the Beta coefficient (β) and the level of ITS-induced SA required to optimise journey length, duration, cost and carbon outputs.

9.4 Summary

"[…] This was beginning to make sense now: AI hadn't replaced human drivers overnight. It had taken years, decades, for automated transport systems such as the Zoomtube, Robochairs, and Mo-pads to be developed and refined in Mega-City One. This slowness to adapt was why you could so often hear Mega-City One's chief transport engineer bemoaning the fact the city would be an engineer's paradise, were it not for the humans."

Contrary to general belief, this exploratory study shows there is a point at which 'more SA', at the individual level at least, does not lead to more efficient road networks at the collective level. Indeed, in many cases the same outcomes can be achieved with 0% SA as they can with 100%. Simply introducing an abundance of ITS into vehicles, and somehow inducing 100% driver SA and compliance, will not necessarily result in the outcomes expected. In the present case the topography of an urban street layout is seen to combine with driver SA to form two important contingency factors for how and when new technology should be deployed. Street patterns and route guidance interact to introduce congestion on the network, define who gets caught in it and create collective effects on time, cost, distance and carbon emissions.

This is necessarily a highly exploratory study that was deliberately kept until the end. Despite this a number of potentially important implications arise. The first relates to previous research which shows between 30% and 50% of drivers do not comply with ITS-based route guidance no matter how much of it is provided. As the results show, middle values like these represent an optimum on many outcome variables and street network types. This raises an intriguing point. Would the city of the future be an engineer's paradise were it not for humans? Or is it precisely 'because' of the humans that it functions at all? Consider this: does the established 30%–50% driver acceptance of route guidance arise through repeated experience of traffic conditions on a network, which itself is a form of strategic SA? If so, could it be taken as evidence of driver behaviour (and the network itself) self-organising over time? If this is the case, is the imposition of ITS based on a false premise? Does 'everyone' in the network need to know 'everything'? The results of this study would seem to suggest that, for some network types, they do not. Moreover, is the collective SA of the drivers in the network responding dynamically to the conditions and finding its own level? Perhaps, in which case the same outcomes could be achieved for considerably reduced cost.

The second implication relates to the different ITS strategies which could be adopted, and when. Is a costly strategy of trying to induce complete SA within drivers optimum in all situations? No. Likewise, is a laissez-faire approach to route planning and route guidance, based purely on ad-hoc local knowledge brought to situations by individual drivers, optimum? Again, no. Optimisation is contingent upon the topology of the network being travelled upon. There are clearly some situations where it would benefit outcome variables such as carbon emissions to enhance tactical forms of SA, and other situations where it would not be useful. The results of this study are helpful in understanding what these relationships might be and what an adaptive tactical feedback strategy might look like. It would be a form of vehicle technology which is cognitively compatible with drivers, one in which the timing and sequence of route guidance information would be oriented around different outcome variables at different times, offering in all cases tangible journey-based 'rewards' for the driver. These rewards would encourage ITS to be used in ways that exceed the current 30%–50% acceptance rate where it is beneficial to do so, and as such, to accumulate some significant marginal gains.

Back in the stygian depths of Mega-City One's undercity, within the abandoned driving simulator laboratory, the professor has the last word: *"Don't you see, the harder we drive the technology, the more we seek to make things logical and machine-like, the more we get all sorts of unexpected problems, which we humans still need to fix. That's the problem with all these dystopian comic-book cities of the future."*

10
CONCLUSIONS

We have crossed the finish line.

We have described our 15-year journey into the human factors of vehicle feedback, linked it to the concept of driver situation awareness (SA), and tried to relate the findings from academic theory to vehicle design practice. For automotive engineers and vehicle dynamists we have endeavoured to present SA as a new margin of human performance, a novel way to design vehicles of the future and a way to access hidden consumer requirements. For human factors readers we have tried to provide a novel perspective on vehicle design, explored and extended the concept of situation awareness and identified new opportunities to constrain undesirable driver behaviours and enhance others. For the reader who did actually own a 1978 Austin Mini, or Porsche 911 Turbo, Infiniti Q50, Citroën CX or the numerous other real-world case study vehicles, we also hope to have provided a human factors basis for why these cars, and all the other countless billions, feel the way they do, and moreover, why those feelings – resulting from the interplay of vehicle feedback and driver SA – are perhaps more important than we think.

We started this research journey knowing anecdotally vehicle feedback influences driver cognition. Vehicle manufacturers are well aware of this 'experientially'. Test drivers routinely put new vehicle prototypes through their paces and report back on how the car felt, often with a high degree of insight, along with what needs to be changed to improve it. On the research side we also know vehicle feedback affects performance in a control-theoretic sense. Indeed, we could already characterise and formalise the driver vehicle interaction with a high level of precision. Even psychophysically we know feedback is readily detectable by drivers and has an influence on their behaviour. What we now know, in addition to all this, is that vehicle feedback also affects driver SA. Establishing a link between SA and driver feedback provides an additional structure in support of experiential knowledge;

it extends and compliments control-theoretic models of the driver/ vehicle system; and it offers an explanation for what drivers do, cognitively, with the feedback psychophysics shows them to be so sensitive to. It seems sensible, therefore, to bring this research monologue to a close by focusing on two key aspects: the implications of this work for the study of situation awareness, and the implications for future vehicle designs.

10.1 Implications for the Study of Situation Awareness

10.1.1 Entering the Mainstream

Vehicle designers may not be aware of this, but SA is one of the most keenly studied topics in the field of Human Factors and also, occasionally, one of its most controversial. As we have seen, the term is used to describe how people, and increasingly entire sociotechnical systems, become and remain coupled to the dynamics of their environment (Moray, 2004). As a concept it provides researchers and practitioners with various models and methods to describe what SA comprises, to determine how individuals, teams or systems develop SA, or to assess the quality of SA during task performance (Salmon and Stanton, 2013). It should provide explanations for what happens when SA is lost, and how it affects performance when it is gained (Stanton et al., 2015). Most importantly in this context, it should provide designers with information to ensure SA requirements are met.

Unusually for a human factors concept the term 'situation awareness' has entered the mainstream lexicon. It is used in many contexts to refer to information residing in people's heads, minds (Fracker, 1991; Sarter and Woods, 1991; Endsley, 1995) or even brains (Endsley, 2015); as something which exists in the world, in displays or other environmental features (e.g. Ackerman, 1998, 2005); as something which is an emergent property of people and their environment (Stanton et al., 2006, 2009a, 2010); or as a form of distributed cognition (e.g. Hutchins, 1995a, b). SA has been explored in countless areas, from military settings to sport, health care and medicine, through to industrial process control and beyond. The papers which deal with the concept are among the top cited in the discipline and the term 'situation awareness' one of the most widely used. It would be surprising if it did not apply in the case of vehicle design. A cursory glance

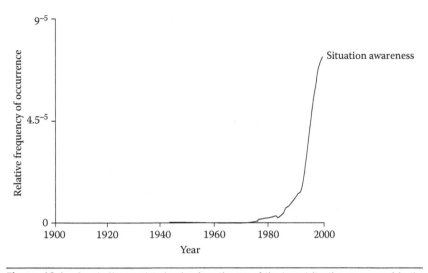

Figure 10.1 Google Ngram plot showing how the use of the term 'situation awareness' in the English language lexicon has accelerated dramatically since the 1980s.

at Google Ngram (Figure 10.1) shows the term 'situation awareness' was hardly used at all within the corpus of English language literature prior to Endsley (1988), Woods (1988) and other pioneering publications on SA in the late 80s, with the line graph accelerating dramatically from that point forward. As the present study of vehicle feedback and driver SA has shown, despite its prevalent use our understanding of SA is still in development (Flach, 1995; Dekker, 2015).

So what, exactly, are the key issues for the study of SA? After all, taken at face value the discipline of Human Factors (HF) has a dominant theory of SA (the three-level model put forward by Endsley in a special issue of *Human Factors* in 1995 is undoubtedly the most popular) and associated methods with which to convert it into HF practice (of which there are lots of examples). If we define the HF discipline purely in terms of a pragmatic philosophy, and judge success only in terms of practical application, then the matter is settled. What makes the study of vehicle feedback and driver SA so interesting, however, is how radically different the concept of SA looks when projected through different worldviews. What is becoming increasingly evident is the worldview through which early models of SA were projected is quite different from the worldview which exists today. We can clearly see this in our own 15-year research journey into the topic of driver feedback. A particularly dramatic case is presented in Chapter 8.

Here we saw how the findings could be interpreted in opposite ways depending on the SA lens used: one lens would lead to the conclusion expert drivers had poorer SA, while another would show they had better SA.

10.1.2 The First and Second 'Cognitive Revolutions'

When Endsley's pioneering paper on SA was published in the 1988 proceedings of the Human Factors and Ergonomics Society's annual meeting in the United States, the dominant paradigm of experimental and cognitive psychology was very much in the ascendency within the discipline. Van Winsen and Dekker (2015) refer to this as the 'first cognitive revolution'. Ideas around systems thinking, specifically distributed cognition (e.g. Hutchins, 1995a,b) and cognitive systems engineering (Rasmussen et al., 1994) were far from mainstream and in many cases in their infancy. Compare this situation to today. There is now a much stronger systems focus (Salmon et al., 2015; Dul et al., 2012) and a growing recognition of the importance of systems concepts. These include concepts such as complexity (Walker et al., 2010), constraints (Vicente, 1999), dynamism (Woods and Dekker, 2000), multiplicity (Lee, 2001), fuzziness (Karwowski, 2000; Lee et al., 2003), randomness (Hancock et al., 2000) and the myriad other terms used to describe the kinds of problems HF practitioners are called upon to help resolve. Nearly 30 years since the term 'situation awareness' entered the HF lexicon the paradigm has shifted, with the second cognitive (systems) revolution, according to Van Winsen and Dekker (2015), upon us. In fact, the paradigm shift is well evident in this book as it traces a journey from individual views of SA (e.g. Chapter 5) to deeply systemic ones (e.g. Chapter 9).

The assumption throughout this book is that SA is important for drivers. Interestingly, however, the relationship between SA and task performance was initially quite difficult to establish (Endsley, 1995). It is unquestionably the case good task performance can occur despite poor SA and vice versa, so clearly the relationship is more complex than a simple one-to-one mapping. Indeed, in Chapter 2 the question was raised about drivers' adaptive capacity given that comparable driving performance was generated on considerably different 'amounts' and 'types' of SA. This has been an important driver for innovations in SA theory, in particular those which advocate a systems SA view. The results are promising. Current distributed SA (DSA) research, of which Chapter 7 and 8 are examples, is providing a much-needed evidence base for how, why and under what conditions SA will lead to improved performance (e.g. Griffin et al., 2010; Sorensen and Stanton, 2013; Sorensen and Stanton, 2016; Rafferty et al., 2013; Walker et al., 2009). These studies suggest DSA is not only a useful concept but also correlated with task success or failure. The issue seems to be that different models of SA are also useful, but perhaps in different circumstances and for different reasons. This, for us, seems to be the key. One (SA) size does not necessarily fit all (research questions).

10.1.3 Contention and Controversy

Recent debates in the SA literature have tended to be somewhat adversarial, with the perceived merits of one approach being compared with the perceived demerits of another (e.g. Endsley, 2015; Salmon et al., 2015; Stanton et al., 2015). Whilst undeniably entertaining and engaging to read, these discipline debates are also genuinely useful. They serve to drive out valid points of issue and subject theories and concepts – no matter how long-standing – to a stress test. Of course, beyond these debates is a more measured picture, one that goes to the heart of wider methodological issues. Simply put there is no 'one best' theory. SA is not a consumer product competing in a marketplace. It depends on the fundamental nature of the problem to which different SA approaches are being applied. All have a role to play. If the practical SA issue being examined can reasonably be characterised as stable, relying on deviations from accepted normative practices, and focusing on individual drivers – as in Chapters 5 and 6 – then there are SA theories which match perfectly to this situation and will deliver the insights needed. If, on the other hand, the problem can be characterised by a socio-technical system in which SA is neither normative or stable, and resides as a systems phenomenon rather than individual one – such as Chapters 7, 8 and 9 – then likewise, other approaches matched to these features will deliver the needed insights. This new 'contingent approach' can be summed up again pictorially in Figure 10.2.

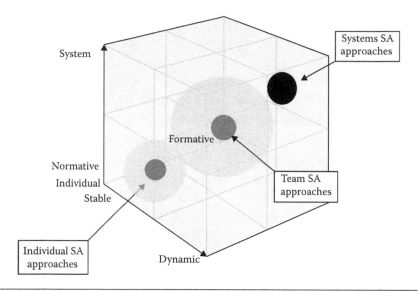

Figure 10.2 Different approaches to SA match to different features of ergonomic problems. By its nature, the more systemic the SA model the more of the problem space it can operate within, although other approaches may prove more practically expedient. The real challenge going forward is to become better at this matching rather than having one approach compete with another.

The key issue for the study of SA is that an overly doctrinaire and rigid approach, not flexible to the nature of the problems being tackled, will (a) likely fail to deliver the required insights, and/or (b) do so with excessive analytical effort compared to alternatives, and/or (c) in the worst case give misleading results.

Despite the sometimes-adversarial nature of discipline debates in this field the future is bright for SA research and practice. Automotive systems continue to become more complex and technology-driven, which in turn raises important questions around awareness and how best to support it across individuals, teams, organisations and entire transportation systems. Critical new SA research questions continue to reveal themselves, such as how drivers can exchange awareness with new vehicle technologies, how we can study SA in teams comprised entirely of non-human agents (such as platoons of autonomous vehicles), and how to design advanced automation systems in which awareness is distributed (e.g. vehicle to vehicle, and vehicle to infrastructure). The challenges and opportunities for SA seem limitless.

10.2 Implications for Future Vehicle Design

10.2.1 The Final Frontier

According to Endsley (2004, p. 337), "The most interesting frontier for SA remains in the design arena". We agree. How can we transform what we know about SA into explicit guidance so that vehicle designers can innovate future designs which enhance, rather than inhibit, SA? Brining this issue into the foreground is timely and important. We saw in Chapter 2 how a reading of current trends could lead one to think modern vehicle design is heading towards greater isolation and reduced driver SA. Taken to its limit one could imagine the kind of 'Zuboff-esque' isolation chamber simulated in Chapter 5, a case in which there is no vehicle feedback at all: no steering feel, no auditory feedback, no tactile feedback. This was a condition put in place as an experimental baseline, yet one only has to consider the Google car, for example, to see it may not be as far-fetched as it seems. An autonomous transport pod – according to strict control theory at least – does not require driver feedback of any sort. It is our experimental baseline brought to life. Faced with such challenges, we hope the vehicle designer might find an overall design process which enables SA requirements to be extracted and met helpful. Such a process begins with the high-level flowchart in Figure 10.3. We invite practitioners to scrutinise their own practice and where these different stages may already be present, or indeed missing.

10.2.2 The SA Design Process

The SA design process begins with an SA requirements analysis (see Chapter 4 and also the materials in the Appendix). This is performed so that all SA requirements, for all different end-users, can be comprehensively identified and recorded. An excellent starting point for an SA requirements analysis – regardless of the theoretical lens being used – is a Hierarchical Task Analysis (HTA). A comprehensive HTA of Driving is provided for this purpose in the Appendix. Templates are also provided which allow analysts to extract the driving tasks under analysis and systematically diagnose what elements need to be perceived, comprehended and projected according to an individual

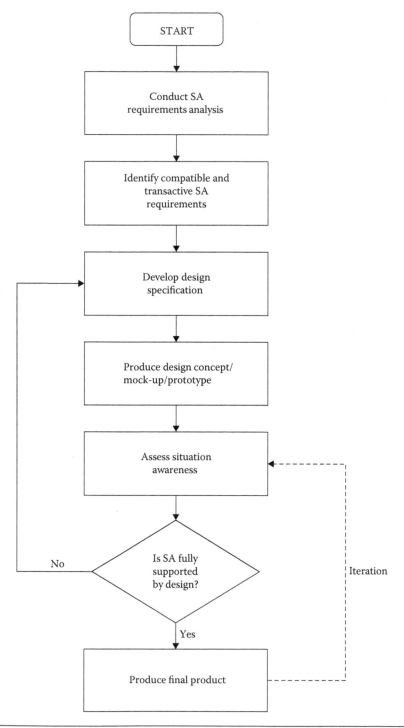

Figure 10.3   SA design process flowchart.

perspective. It also allows 'ownership' of different elements to be identified, and also the means by which shared elements are mediated between different agents in the system, and whether those means are appropriate. This reflects a more team- or systems-based view. It is important to note the SA requirements analysis phase (as described in detail in Chapter 4) does not involve merely identifying the different pieces of information which need to be known; rather, it involves going further and identifying what it is that needs to be known, how this information is used, and what the relationships between the different pieces of information actually are. Identifying the relationships between different pieces of information allows designers to group information meaningfully in the end design. For a comprehensive 'multi-perspective' view of SA we recommend using the outputs of the HTA-based requirements analysis to construct information networks of the sort presented in Chapters 7 and 8.

Following the SA requirements analysis phase it is next important to identify which of the information elements represent compatible SA information elements (i.e. those used in different ways by different agents in the system), which are transactive SA information elements (i.e. passed between agents), and which are both. This involves taking the SA requirements analysis outputs (i.e. the HTA and information networks) and, in conjunction with subject matter experts and other investigation (e.g. observation, experimentation etc.) classifying each information element accordingly. A template is provided in the Appendix to enable readers to do this.

The SA requirements analysis outputs and the compatible and transactive SA elements classification should then be used to inform the development of an SA oriented design specification. Again, it is important to note this should not involve merely specifying what it is the system should be presenting to different agents. Rather, this should involve a specification of what it is the 'driving system' (in the wider sense) should be presenting to whom, in what format, what other information should be presented in conjunction (i.e. the relationships between different classes of information) and also what the information is going to be used for (which may be many different things depending on who is using it).

Based on the design specification, mock-ups should then be developed after which SA testing should begin. It is critical for SA to be tested throughout the design lifecycle if SA requirements are to be supported. The assessment involves determining the extent to which the design supports the SA requirements specified, although the exact nature of the assessment is dependent on the stage the design has reached. For example, at the mock-up stage, the assessment might entail walking through the task with subject matter experts and evaluating the extent to which SA requirements, and the relationships between SA-related information elements, are manifest in the design in question. On the other hand, when the design is at a prototype stage, the assessment could involve assessing user performance and SA during actual operational trials. There is a wide range of SA assessment methods which can be used in this phase, and the interested reader is referred to Stanton et al. (2013) for detailed guidance.

Any SA-related difficulties found during SA testing send the design back into the iterative phase of developing specifications, producing concepts and mock-ups, and then testing them again. This cycle also continues through mock-up, concept and prototype versions of the system. Only when designers are satisfied that SA requirements are fully supported can the prototype design proceed from this cycle into the development of the final design.

10.2.3 SA Design Principles

What, exactly, are the outcomes of the SA design process? Obviously, the specific outcomes will differ with each individual design project but there are some broader principles which can be extracted. Before presenting these it is perhaps worthwhile to recap on what a principle is and the purpose it should serve. To do this we can refer to the literature on sociotechnical systems in which the goal to 'jointly optimise' people and technology is shared with Human Factors.

To clarify, a principle is a proposition which serves as a foundation for a system of belief or behaviour or for a chain of reasoning. Good principles should be succinct and clear, and make it explicit what the desired outcome should be and who should action them. The primary benefit of SA design principles is to help embed a set of human-centred values in systems which are typically viewed from a purely technological or engineering perspective (e.g. Badham et al., 2000;

Majchrzak and Borys, 2001). What principles 'are not' are simple prescriptions for action, rules to adhere to rigidly, or a recipe to be enacted without deviation. They are "guides to critical evaluation of design alternatives making clear some of the differences between the sociotechnical systems approach and traditional [system] design" (Berniker, 1996, p. 19). Principles, therefore, "...are not intended as design rules for mechanistic application. Rather, they provide inputs to people working in different roles and from different disciplines who are engaged collaboratively in design. They offer ideas for debate, providing rhetorical devices through which detailed design discussions can be opened up and elaborated" (Clegg, 2000, p. 474). Fourteen SA design principles are presented below. In a sense they represent 'good Human Factors practice', and perhaps at first glance seem a little obvious. So to tease out the implications of these principles their 'antithesis' is presented. It then becomes clear what value an SA-guided design can provide.

Principle 1: Clearly define and specify SA requirements

The importance of knowing what it is that different users 'need to know' during driving task performance has been shown throughout the research presented in this book. The collaborative vehicle design process should, therefore, begin with a clear definition and specification of the SA requirements of the overall driver/vehicle/road system. This should include a breakdown of both compatible and transactive SA requirements. As described above, it is important that SA requirements specification includes more than just the different pieces of information which need to be known, and should go further to describe what it is that needs to be known and by whom, how this information is used by different users, and what the relationship between the different pieces of information actually are, in other words, how they are integrated and used by different users. Matthews et al. (2004) point out that knowing what the SA requirements are for a given domain provides engineers and technology developers with a basis upon which to develop optimal vehicle designs to maximise driver performance. It is recommended SA requirements analysis be conducted using Hierarchical Task Analysis (a completed analysis appears in the Appendix) combined with the information network approach described in Chapters 7 and 8.

ANTITHESIS

The SA requirements of drivers are simply assumed, based on experiential 'common sense' knowledge of the current situation, or missed completely. After all, the vehicle itself does not need the feedback provided to the driver (e.g. steering feel, engine noise etc.) and the fact it proves so useful has arisen due to product design evolution over a long period. We currently do not know the adaptive limits of drivers when their SA is reduced. The design risk is that SA requirements are not met, leading to a wide variety of errors and wider system pathologies.

Principle 2: Ensure 'agent' roles within the wider 'system' are clearly defined
The driver/vehicle/road system consists of multiple members, each of which have distinct goals, roles, and responsibilities. The different roles and responsibilities within this 'driving system' need to be clearly defined. The word 'agent' is used deliberately. SA does not reside purely in the minds of humans, but is distributed in various ways across the wider system in which the human, and other 'agents' which perceive, store, display or otherwise use information, are a part.

ANTITHESIS

A highly local and specific view is taken in which feedback is arbitrarily removed from parts of the driving system which do not require it, but with little thought given to the knock-on effects in other parts of the system. For example, the vehicle alone does not need the feedback perceived by drivers, but as an 'agent' in a wider system of human actors, infrastructure, technology etc. it has a role to play (at the very least) in mediating information flow which other system agents can use.

Principle 3: Support meta-SA through training, procedures and displays
Meta-SA, that is, awareness of what other system agents know, has been found to be important because it enables SA transactions

to occur at the appropriate time between appropriate system agents. Meta-SA should, therefore, be supported. This can occur through the use of training (as shown in Chapter 8 for example) but it could also be reflected in procedures (e.g. new requirements for highway legislation perhaps?) and, of course, display design. The guiding principle active here is the importance of system agents having an understanding of what other agents are doing and therefore should 'know' at different times during driving task performance; this allows system agents to understand when and where information is required. Stanton et al. (2006), for example, point out that "it is important for the agents within a system to have awareness of which agent is likely to hold specific views and, consequently, to interpret the potential usefulness of information that can be passed through the network in terms of these views" (p. 1308). Further, they go on to point out there are two aspects of SA at any given node in a distributed team: the individual SA of one's own task and the 'meta' SA of the entire system's SA.

ANTITHESIS

Drivers have poor self-awareness of their own SA and therefore run the risk of 'losing the picture' and not having 'meta-SA'.

Principle 4: Design to support compatible SA requirements

'Agents' in the complete 'driving system' each have distinct but compatible SA requirements. Vehicles should therefore be designed to cater for these compatible SA requirements, rather than to simply support 'shared SA'. In other words, rather than present 'everything' to 'everyone', SA-optimised systems should be designed so that users are not presented with information, tools and functionality they do not explicitly require. Vehicle systems should, therefore, be designed to support the roles, goals and SA requirements of each agent at the appropriate time. This might conceivably involve different displays, tools, functions and forms of feedback for the different roles and tasks involved, or might involve the use of customisable and/or context-aware interfaces and displays.

ANTITHESIS

Information overload: a vehicle interior festooned with buttons and displays in which every conceivable information requirement is presented simultaneously.

Principle 5: Design to support SA transactions

Systems and interfaces which present information to drivers should be designed to support SA transactions where possible. This might involve presenting incoming SA transaction information in conjunction with other relevant information (i.e. information about what the incoming information is related to and is to be combined with) and also providing users with clear and efficient communication links with other agents in the driving system where this is necessary.

ANTITHESIS

Arbitrary and disconnected information is presented to the driver with no support for how that information is actually useful.

Principle 6: Group information based on links between information elements

The network-based approach to SA described in Chapters 7 and 8 help reveal which elements of information are used together for SA acquisition and maintenance at different points in the driving task. Vehicle designers can group such information (on interfaces and displays for example) based on the links between information elements specified in the requirements analysis.

ANTITHESIS

Information is presented to the driver in an ungrouped fashion. Different components may occur in different interfaces, via different modes, at different times, making it difficult for drivers to integrate.

Principle 7: Remove unwanted information

While we have dealt at length with cases of 'feedback isolation' (e.g. see Chapter 5) the opposite also holds true. Information not needed by drivers should generally not be presented. Driver displays should be designed so that, based on SA requirements, unwanted information can be removed or hidden. The information needed by one agent in the system (e.g. the driver) may be very different to that needed by another agent (e.g. other drivers, the vehicle etc.).

ANTITHESIS

An overly complex and unintuitive dashboard/driver display which is difficult to use, time consuming and at odds with the driver's needs and expectations.

Principle 8: Consider the technological capability available and its impact on SA

The work presented in this book highlights the problems associated with technology limitations that can degrade SA. Again, perhaps an obvious, but nevertheless critical recommendation is system designers need to carefully consider the constraints imposed on them by technological capability and design the system accordingly within these constraints. SA requirements may need more, or indeed less, technological capability than is currently available. SA provides a means to target where particular technology is needed and what its capabilities should be. SA requirements create the possibility to direct technology into delivering user benefits without the need for superfluous or otherwise demanding technical requirements. In other words, the user sets the information needs rather than the capabilities (or limitations) of technology.

ANTITHESIS

Expensive vehicle technology which despite its impressive appearance is actually at an 'intermediate level of intelligence'. Sufficient to support some, but not all SA needs.

Principle 9: Ensure information presented to users is accurate at all times

The transmission of erroneous information can lead to erroneous SA and a reduction in performance and satisfaction. The information presented by a vehicle system should be accurate.

ANTITHESIS

Direct feedback and interaction is replaced by poorly designed 'indirect' forms of interaction, drive-by-wire systems which introduce inaccuracies in control feel, for example.

Principle 10: Ensure information is presented to drivers in a timely fashion

Information not presented in a timely manner can lead to SA decrements. SA-related information should be presented to users without delay. Further, the timeliness of information should also be represented on interfaces and displays where possible, allowing users to determine information latency.

ANTITHESIS

Immediate feedback and interaction is replaced by 'laggy' forms of indirect interaction, sluggish controls and interfaces which do not keep pace with the driving environment, for example.

Principle 11: More information is not always better

Users should be presented only with information they specifically require rather than all of the information the system is capable of presenting. According to Bolia et al. (2007), an increased amount of available information does not necessarily mean users of the data will make better decisions.

ANTITHESIS

Information overload. The sheer quantity of information presented to drivers is excessive and poorly integrated and grouped, with no clear priority given to what is more or less important.

Principle 12: Present SA-related information in an appropriate format

Drivers are required to assimilate and understand large volumes of information. It is critical that systems are designed so information presented to users is in a format amenable to quick, efficient and accurate assimilation and understanding.

ANTITHESIS

The already-loaded visual modality is loaded further still by information which could (and previously was) communicated by implicit vehicle feedback cues, for example, head-up speed displays to make up for the absence of engine noise, G-force readings instead of steering feel etc.

Principle 13: Use procedures to facilitate SA

Chapter 8 indicates that procedures and training are an effective means of facilitating SA acquisition and maintenance via SA transactions. Procedures, scripts or routines taught via established driver training methods could help new drivers to know where to look for information and what to do with it. Likewise, new technologies could conceivably guide behaviours towards those advocated by advanced driver training – perhaps some form of advanced driving co-pilot? – helping to foreground the acquisition of information and enabling appropriate driving behaviours based on it.

ANTITHESIS

A generation of drivers who do not know where to look for information to support their driving, or what to do with it. A similar generation of drivers who are unaware of what future technology is doing to the driving task and not able to benefit from potential enhancements.

Principle 14: Test SA throughout the system lifecycle

It is clear SA should be considered and tested where possible throughout the design lifecycle. Despite this, from our experience this is not always the case. Formal SA assessments are comparatively rare in the automotive sphere, and often only undertaken as an afterthought once systems are operational and suggested redesigns are too expensive to implement. It is recommended, therefore, that SA is tested throughout the system design lifecycle, from the concept phase, through the mock-up and prototype phases, to the operational end product. This might involve simply assessing DSA using functional diagrams of an interface and subject matter experts, or it might involve large-scale operational trials of prototypical systems.

ANTITHESIS

A car design in which fundamental deficiencies in SA continue to propagate and which is unresponsive to the changing needs and requirements of the driver and the environment in which the vehicle is being operated.

10.2.4 Principles in Practice

We began this monograph with an amusing comparison between older and newer vehicles. There can be no doubt at all that modern cars are vastly superior in virtually all technical aspects than vehicles of thirty years previous. We are emphatically not recommending a return to the cars of yesteryear. What we are suggesting, however, is that valuable user-interaction lessons have been acquired over the previous 150 years of automotive evolution, lessons which could be useful for the next generation of vehicles, and which in some important respects risk being forgotten. Let us start to draw this monograph to a close by putting these notions in contact with two vehicles which share the same name but are separated by five decades of automotive design evolution, as shown in Figure 10.4.

The SA design principles are clearly manifest in the vehicles which fill our streets, but the study of SA – we believe – does more than that. It also connects with the more ephemeral nature of cars and driving, something we have deliberately incorporated into the writing style of this book. Consider this. In purely utilitarian terms cars should really be painted in black-and-yellow hazard stripes, because that is

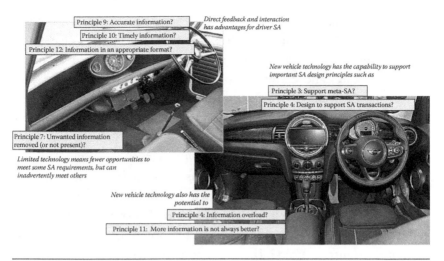

Figure 10.4 SA design principles in practice.

the safest and most visible colour scheme; they would mostly be small single-seater vehicles because most cars, most of the time, carry only one occupant; and they would not be capable of more than approximately 20 mph because in urban areas why is there a need to go faster? As a purchase – if indeed 'purchase' is the future business model – cars would be much more like a washing machine. A 'white good'. A simple utility. Framed in these terms there is an argument to say current cars, therefore, are dramatically oversized, overengined and overengineered. And they need to be, because purely utilitarian vehicles do not succeed, at least in the consumer-driven Western market which foregrounds ephemeral aspects such as social status, notions of luxury and performance, and feelings of safety, but also fun and driver enjoyment. Herein lies the paradox.

Looking into the future as every large car manufacturer is currently doing, some significant trends are clearly discernible. One of them is the notion of 'transport as a service'. This includes autonomous vehicles, ride sharing, new ownership models, new powertrains and, underneath all this, a requirement to either support people in what they already do (but in new ways) or else for people to change their behaviour completely. Looking at Ford's vision of the future (www.youtube.com/watch?v=FO824cwTYJY), in which any similarity to 'Mega-City One' (Chapter 9) is purely coincidental (!), one can see a re-prioritisation of the urban landscape to favour active travel, with reduced road capacity made up for by new technology. Looking closely, some of the vehicles on this imagined network appear to be precisely the small, single-seater vehicles a purely utilitarian approach would advocate, and certainly not the oversized, overengined, and overengineered vehicles we know (and purchase) today. The question is how do we get from today's vehicles to this future vision, assuming this is what we want? The challenges are formidable. The technical challenges of vehicle autonomy and new powertrains are challenging enough but an equal, if not greater challenge is behaviour change. How can drivers be persuaded to switch from a Ford Explorer or Focus, let alone a 'high feedback' GT40 or Mustang, into a semi-autonomous, electrically powered 'mobility solution', possibly shared with other transport users? This is far from a trivial point.

What this large-scale and currently imagined shift requires is an even greater focus on the user and their needs, and an even greater focus on the wider system in which transport solutions reside. On these terms the Human Factors concept of situation awareness could be an important enabler for behaviour change. Put it this way: what if the purely utilitarian vehicle of the future, which we might need, can be made to 'feel like' the overengineered vehicle of today with its ephemeral notions of status, enjoyment and driver engagement consumers seem to want? The research presented in this monograph shows that drivers are very sensitive to how their vehicles feel, and the changes to be wrought via new drive-by-wire and powertrain technologies are orders of magnitude greater than what drivers are capable of detecting. The opportunity inherent in this technology, which is available for vehicle designers to harness, is that new 'mobility solutions' could be made to feel even better to drive than ever. The desirable vehicle choice could also be the cleanest, most journey-appropriate, sustainable one. The possibilities for future vehicles, and future driver–vehicle interactions, are considerable.

Appendix

This is the complete Goal-Directed (Hierarchical) Task Analysis of Driving developed by the authors for the purposes of defining SA requirements. It comprises over 1600 bottom-level tasks and 400 plans and is based on the following documents, materials and research:

- The task analysis conducted in 1970 by McKnight and Adams
- The latest edition of the UK Highway Code
- UK Driving Standards Agency information and materials
- Mares et al.'s (2013) *Roadcraft* (the Police Foundation/Institute of Advanced Motorists driver's manual)
- Subject-matter expert input (such as police drivers) originating from our research in advanced driver training
- Numerous on-road observation studies involving a broad cross-section of 'normal' drivers

The Hierarchical Task Analysis of Driving (HTAoD) models the task normatively as it should occur in the UK. There will be many similarities to other nations, but please note the UK drives on the left-hand side of the road. This means, for example, right-hand turns conflict with opposing traffic flows instead of the other way around; we sit on the right-hand side of the car to drive it; and, as modelled here, vehicles with manual gearboxes are the norm.

The analysis has proceeded through numerous validation and modification phases since being originally developed in 2001, but like any similar HTA there will be different ways to arrive at a broadly similar set of bottom-level tasks. It is offered here for researchers to use and modify as they see fit, and as a contribution to the wider field of transportation human factors. The only stipulation is that acknowledgement using the following citation is made in cases where the analysis is used or repurposed:

Walker, G. H., Stanton, N. A., and Salmon, P. M. (2018). *Vehicle Feedback and Driver Situation Awareness.* Boca Raton, FL: CRC Press.

A tabular format is used with full notation. Line indentations denote different HTA levels and distinguish between super- and subordinate goals. Plans are constructed using logical operators (such as AND, IF, THEN, ELSE, WHILE etc.). To provide a more space-efficient way of presenting this complex task, and to avoid the need to keep reproducing recurring clusters of behaviour, the analyst is referred to other parts of the analysis using the notation 'GO TO subroutine'. This is not intended to represent a break with the theoretical underpinnings of HTA. The higher-level plans still capture the parallelism and sequencing of lower-level goals/tasks; the GO TO subroutine method is not intended to replace this, merely to avoid repetition. For completeness, analysts could use the GO TO prompts to reproduce the task clusters in full if it suited their needs better.

Also included, at the end of the HTA itself, are some SA requirements templates. These can be populated with relevant driving tasks/goals, and will help the analyst define the SA requirements required therein.

TASK ANALYSIS OF DRIVING - SUMMARY

0 TASK STATEMENT
1 PERFORM PRE-DRIVE TASKS
   1.1 perform pre-operative procedures
   1.2 start vehicle
2 PERFORM BASIC VEHICLE CONTROL TASKS
   2.1 pull away from standstill
   2.2 perform steering actions
   2.3 control vehicle speed

2.4 decrease vehicle speed

2.5 undertake directional control

2.6 negotiate bends

2.7 negotiate gradients

2.8 reverse the vehicle

3 PERFORM OPERATIONAL DRIVING TASKS

3.1 emerge into traffic from side road

3.2 follow other vehicles

3.3 overtake other moving vehicles

3.4 approach junctions

3.5 deal with junctions

3.6 deal with crossings

3.7 leave junction (crossing)

4 PERFORM TACTICAL DRIVING TASKS

4.1 deal with different road types/classifications

4.2 deal with roadway-related hazards

4.3 react to other traffic

4.4 perform emergency manoeuvres

5 PERFORM STRATEGIC DRIVING TASKS

5.1 perform surveillance

5.2 perform navigation

5.3 comply with rules

5.4 respond to environmental conditions

5.5 perform IAM system of car control

5.6 exhibit vehicle/mechanical sympathy

5.7 exhibit driver attitude/deportment

6 PERFORM POST-DRIVE TASKS

6.1 park the vehicle

6.2 make the vehicle safe

6.3 leave the vehicle

THE TASK: Drive a car

THE CONTEXT: A modern, medium-sized UK specification* front wheel drive vehicle on a UK (left-hand drive) public road

THE PERFORMANCE CRITERIA: Drive in compliance with the Highway Code, and the Police Driver's System of Car Control

* Specification includes a manual gearbox with floor-mounted gear lever, a manual hand brake (also with floor mounted lever), and a fuel-injected engine.

0 TASK STATEMENT

Drive a modern, average sized, front wheel drive vehicle equipped with a fuel-injected engine, on a British public road, in compliance with the Highway Code, using the Police Driver's System of Car Control.
Plan 0 - do 1 THEN 2 AND 3 AND 4 WHILE 5 THEN 6

1: PERFORM PRE-DRIVE TASKS

Plan 1 - do 1 THEN 2
1.1 perform pre-operative procedures
Plan 1.1 do in order
   1.1.1 enter car
   Plan 1.1.1 - do in order
      1.1.1.1 unlock door
      1.1.1.2 open door
      1.1.1.3 sit inside car
      1.1.1.4 shut door
   1.1.2 perform pre-drive checks
   Plan 1.1.2 - do in order
      1.1.2.1 check vehicle status
      Plan 1.1.2.1 - do 1 THEN 2
         1.1.2.1.1 check handbrake is on
         1.1.2.1.2 check gearstick is in neutral position
      1.1.2.2 check and adjust seating preferences
      Plan 1.1.2.2 - do 1, IF adjustment to driving position is needed
             do 2 AND/OR 3 AND/OR 4
         1.1.2.2.1 check seating position
         1.1.2.2.2 adjust longitudinal position
         1.1.2.2.3 adjust backrest
         1.1.2.2.4 adjust head restraint
      1.1.2.3 check and adjust mirrors
      Plan 1.1.2.3 - do 1, IF adjustment is needed do 2 AND/OR 3
         1.1.2.3.1 check mirror positions

       1.1.2.3.2 adjust side mirrors

       1.1.2.3.3 adjust rear-view mirror

    1.1.3 put on seatbelt

1.2 start vehicle

Plan 1.2 - do in order

    1.2.1 use ignition key

    Plan 1.2.1 - do in order

       1.2.1.1 put key into ignition switch barrel

       1.2.1.2 turn ignition key to position 1

       1.2.1.3 release steering lock

    1.2.2 prepare to start engine

    Plan 1.2.2 - WHILE 6 do 1,2,3,4,5

       1.2.2.1 switch off unnecessary electrical systems

       1.2.2.2 depress clutch pedal

       1.2.2.3 recheck gearbox is in neutral

       1.2.2.4 turn ignition key to position 2

       1.2.2.5 wait for fuel injection dashboard light to extinguish

       1.2.2.6 do not depress accelerator

    1.2.3 start engine

    Plan 1.2.3 - do 1 THEN 2 AND 3. IF engine fires do 5. IF engine
                doesn't fire do 4, THEN 5 THEN 6 THEN repeat
                plan. IF engine does not start after 8 attempts OR
                the battery is starting to run flat THEN 7

       1.2.3.1 turn ignition key to position 3

       1.2.3.2 hold key in position 3

       1.2.3.3 check if engine is starting to fire

       1.2.3.4 hold key in position 3 for 15 seconds

       1.2.3.5 release key (reverts to position 2)

       1.2.3.6 wait for 30 seconds

       1.2.3.7 abandon drive/seek assistance

    1.2.4 raise clutch pedal slowly

    1.2.5 check warning lights and engine status

    Plan 1.2.5 - do simultaneously

       1.2.5.1 check ignition and oil warning lights

       Plan 1.2.5.1 - IF warning lights remain illuminated for <
                    5 seconds THEN 1. IF warning lights still
                    illuminated THEN 2. IF warning lights still
                    illuminated THEN 3

  1.2.5.1.1 blip throttle

  1.2.5.1.2 rev engine to 3000RPM for 4 seconds

  1.2.5.1.3 abandon drive and seek assistance

 1.2.5.2 check airbag warning light

Plan 1.2.5.2 - IF light is on THEN 1 ELSE 2

  1.2.5.2.1 wait 30 seconds

  1.2.5.2.2 abandon drive and seek assistance

 1.2.5.3 check engine status

Plan 1.2.5.3 - WHILE 3 IF engine not ticking over at between 800 and 1200RPM THEN 2. IF engine emitting abnormal sounds THEN 4

  1.2.5.3.1 observe rev counter

  1.2.5.3.2 blip accelerator

  1.2.5.3.3 check engine sounds

  1.2.5.3.4 abandon drive and seek assistance

1.2.6 operate in car systems

Plan 1.2.6 - 1 AND/OR 2 AND/OR 3 as required/desired

 1.2.6.1 operate interior heating ventilation

 1.2.6.2 operate radio

 1.2.6.3 operate other in-car devices

2: PERFORM BASIC VEHICLE CONTROL TASKS

Plan 2 - IF pulling away from standstill THEN 1 ELSE 2 AND 3 AND/OR 4 AND/OR 5 AND/OR 6 AND/OR 7 AND/OR 8 as required

2.1 pull away from standstill

Plan 2.1 - IF pulling away on a gradient THEN 2 IF pulling away on slippery surface THEN 3 ELSE 1

 2.1.1 set off on level ground

 Plan 2.1.1 - do steps 1,2,3 in order

  2.1.1.1 set controls ready for pulling away

  Plan 2.1.1.1 - do 1,2,3. IF gear lever unwilling to select 1st gear THEN 4, 5 AND repeat plan at step 2.

   2.1.1.1.1 place left hand on gear stick

   2.1.1.1.2 depress clutch pedal fully with left foot

   2.1.1.1.3 move gear lever from neutral position to first gear position with left hand

2.1.1.1.4 move gear lever back into neutral position

2.1.1.1.5 raise clutch pedal fully

2.1.1.2 operate controls in order to initiate manoeuvre

Plan 2.1.1.2 - WHILE 1 do 2. IF engine note changes THEN 3 for no longer than 30 seconds WHILE 4 as required. IF longer than 30 seconds THEN 5 AND 6. IF ready to initiate pulling away manoeuvre THEN repeat at step 1. IF ready/safe to pull away THEN 7 AND 8 THEN 9 AND 10. IF engine faltering THEN 9

2.1.1.2.1 using right foot depress accelerator (slightly)

2.1.1.2.2 raise clutch pedal (slowly)

2.1.1.2.3 hold clutch pedal still in position

2.1.1.2.4 perform observation checks as required << GO TO subroutine 5.1 'surveillance' >>

2.1.1.2.5 depress clutch

2.1.1.2.6 release pressure on accelerator pedal

2.1.1.2.7 raise clutch pedal a little more

2.1.1.2.8 release handbrake

Plan 2.1.1.2.8 - do in order

2.1.1.2.8.1 place left hand on handbrake lever

2.1.1.2.8.2 pull lever up (slightly)

2.1.1.2.8.3 depress knob

2.1.1.2.8.4 lower lever to the floor

2.1.1.2.8.5 release hand from lever

2.1.1.2.9 press accelerator further down (gradually)

2.1.1.2.10 let the clutch pedal come right up (smoothly)

2.1.1.3 complete pulling away manoeuvre

Plan 2.1.1.3 - WHILE 2 do 1

2.1.1.3.1 remove foot from clutch pedal

2.1.1.3.2 depress accelerator pedal an amount proportional to desired rate of acceleration

2.1.2 set off on gradient

Plan 2.1.2 - IF gradient is uphill THEN 1. IF gradient is downhill THEN 2

2.1.2.1 pull away on uphill gradient

Plan 2.1.2.1 - do 1,2,3 (smoothly/progressively/briskly)

2.1.2.1.1 prepare to set off on uphill gradient

Plan 2.1.2.1.1 - do 1,2,3 IF engine note changes THEN 4 for 30 seconds THEN 5

    2.1.2.1.1.1 increase engine speed to around 2000/3000RPM

    2.1.2.1.1.2 maintain engine at this speed

    2.1.2.1.1.3 raise clutch

    2.1.2.1.1.4 hold clutch still in this position

    2.1.2.1.1.5 perform observation checks as required << GO TO subroutine 5.1 'surveillance' >>

  2.1.2.1.2 hold vehicle on uphill gradient using engine and clutch

Plan 2.1.2.1.2 - WHILE 4 do 1,2,3. WHILE 6 do 5

    2.1.2.1.2.1 place left hand on handbrake lever

    2.1.2.1.2.2 lift lever

    2.1.2.1.2.3 depress and hold lever knob down

    2.1.2.1.2.4 depress accelerator pedal down a little more

    2.1.2.1.2.5 raise clutch pedal a little more

    2.1.2.1.2.6 release handbrake smoothly

2.1.2.1.3 pull away on uphill gradient

Plan 2.1.2.1.3 - do 1. IF speed <6mph THEN 2 AND 3 THEN 4

    2.1.2.1.3.1 press accelerator a little more (gradually)

    2.1.2.1.3.2 release clutch (smoothly)

    2.1.2.1.3.3 depress accelerator an amount proportional to desired rate of acceleration

    2.1.2.1.3.4 remove left foot from clutch pedal

2.1.2.2 pull away on downhill gradient

Plan 2.1.2.2 - do 1 as required. [perform steps 2,3,4,5,6 briskly] do 2 IF car is felt to tug slightly against handbrake THEN 3,4 AND 5. IF vehicle speed <6mph THEN 6 WHILE 7

    2.1.2.2.1 perform observation checks << GO TO subroutine 5.1 'surveillance' >>

    2.1.2.2.2 start raising clutch pedal with left foot

    2.1.2.2.3 release handbrake << GO TO subroutine 2.1.1.2.8 'release handbrake' >>

    2.1.2.2.4 depress accelerator (gently)

    2.1.2.2.5 release clutch pedal all the way (gradually)

2.1.2.2.6 remove left foot from clutch pedal

2.1.2.2.7 operate accelerator

2.1.3 set off on slippery road surface (e.g. snow or ice)

Plan 2.1.3 - WHILE 4 do steps 1,2,3 in order

    2.1.3.1 prepare to set off on slippery surface

    Plan 2.1.3.1 - do in order

        2.1.3.1.1 straighten front wheels

        2.1.3.1.2 put car in 2nd gear << GO TO subroutine 2.1.1.1. 'set controls ready for pulling away' - at step 3, 'move gear lever from neutral position to 2nd gear position' >>

        2.1.3.1.3 perform observation checks as required << GO TO subroutine 5.1 'surveillance' >>

    2.1.3.2 begin setting off on slippery surface procedure

    Plan 2.1.3.2 - IF safe and/or appropriate to pull away THEN 1 AND 2. IF engine note changes THEN 3,4. do 5 until speed <8mph

        2.1.3.2.1 raise clutch (very gradually)

        2.1.3.2.2 depress accelerator (very gradually)

        2.1.3.2.3 release handbrake << GO TO subroutine 2.1.1.2.8 'release handbrake' >>

        2.1.3.2.4 place left hand back onto steering wheel

        2.1.3.2.5 slip clutch (smoothly)

        Plan 2.1.3.2.5 - do 1 AND 2

            2.1.3.2.5.1 depress accelerator (slightly)

            2.1.3.2.5.2 hold clutch in position

    2.1.3.3 complete setting off on slippery surface procedure

    Plan 2.1.3.3 - IF speed <8mph AND vehicle safely in motion with no wheel spin or loss of grip THEN do 1,2 WHILE 3 ELSE 4 THEN 5

        2.1.3.3.1 fully release clutch

        2.1.3.3.2 remove left foot from clutch pedal

        2.1.3.3.3 depress accelerator (very gently) to obtain a very gentle rate of acceleration

        2.1.3.3.4 depress clutch fully

        2.1.3.3.5 << GO TO subroutine 2.1.3.2 'begin setting off on slippery surface procedure' >>

    2.1.3.4 avoid jerky or harsh control inputs

2.2 perform steering actions

Plan 2.2 - WHILE 2 AND 7 do 1. IF steering manoeuvre demands THEN 3. IF normal driving THEN 4. IF extreme manoeuvre OR rate of turn not brisk enough THEN 5. IF reversing THEN 6 as necessary. IF braking (firmly)/cornering/driving through deep surface water THEN 3 AND 8. IF driving in conditions of low friction THEN 9

   2.2.1 hold wheel lightly

   2.2.2 position hands at quarter to 3 on steering wheel rim

   2.2.3 grip wheel firmly

   2.2.4 pull-push steering method [left/right turn]

Plan 2.2.4 - WHILE 1, do 2,3 AND 4,5 AND 6 in order. repeat steps 2,3 AND 4,5 AND 6 as required to complete turn. IF returning to straight ahead position THEN 7 WHILE 8

   2.2.4.1 do not allow either hand position to go past 12 o'clock on steering wheel rim

   2.2.4.2 slide left/right hand up to a higher position on wheel rim

   2.2.4.3 pull the wheel down with this hand

   2.2.4.4 slide other hand down, keeping level with pulling hand until it nears the bottom of the wheel

   2.2.4.5 start pushing up with the other hand

   2.2.4.6 slide pulling hand up the wheel, keeping level with pushing hand

   2.2.4.7 feed wheel back through hands with opposite movements described in steps 2,3 AND 4,5 AND 6

   2.2.4.8 do not permit self-centring action to spin wheel on its own

   2.2.5 perform rotational steering

Plan 2.2.5 - IF turn >120degrees THEN 1. IF turn of <120degrees required THEN 2,3,4. repeat steps 2,3,4 as required to complete turn. IF turn of <120degrees is anticipated THEN 5. IF returning to straight ahead position THEN 6 WHILE 7

   2.2.5.1 turn wheel with light but fixed hand hold at 1/4 to 3

   2.2.5.2 reposition lower hand at 12 o'clock (cross arms as required)

2.2.5.3 use lower hand to continue (smoothly) pulling down the wheel

2.2.5.4 place other hand near the top of the wheel to continue the turning motion

2.2.5.5 place leading hand at top of wheel rim before starting turn

2.2.5.6 use a similar series of movements but in the opposite direction

2.2.5.7 do not permit self-centring action to spin wheel on its own

2.2.6 perform reversing hold

Plan 2.2.6 - WHILE 3 do 1 AND 2 as necessary to complete turn

2.2.6.1 put one hand on top of the steering wheel rim

2.2.6.2 control movement of steering wheel with this hand

2.2.6.3 use the other hand to hold the wheel low down

Plan 2.2.6.3 - IF allowing wheel to pass through hand as other hand operates steering wheel THEN 1. IF holding wheel in position whilst other hand is being repositioned THEN 2

2.2.6.3.1 hold loosely

2.2.6.3.2 grip tightly

2.2.7 avoid turning steering wheel while vehicle is stationary

2.2.8 keep both hands on steering wheel

2.2.9 steer as delicately as possible

2.3 control vehicle speed

Plan 2.3 - WHILE 5 IF accelerating away from rest THEN 1 IF increasing current speed to new target speed THEN 2 IF maintaining existing road speed THEN 3 IF extremely low speed is required THEN 4

2.3.1 accelerate from rest

Plan 2.3.1 - do 1 AND 2

2.3.1.1 use accelerator pedal to control engine power delivery

Plan 2.3.1.1 - do 1. IF normal acceleration demanded THEN 2. IF brisk acceleration desired THEN 3. IF full power acceleration demanded THEN 4 (WHILE using 1st OR 2nd OR 3rd gear(s) only)

2.3.1.1.1 depress accelerator pedal an amount proportional to desired rate of acceleration

2.3.1.1.2 allow engine revs to rise to approx. 3500RPM before upshifting

2.3.1.1.3 allow engine revs to rise to approx. 5000RPM before upshifting

2.3.1.1.4 allow engine revs to rise to 6500RPM before upshifting

2.3.1.2 change up gears

Plan 2.3.1.2 - do 1. WHILE 2 do 3 THEN 4. WHILE 5 attempt to match engine speed with road speed by performing step 6. WHILE 7 do 8,9,10 in order

2.3.1.2.1 place left hand on gear stick

2.3.1.2.2 depress clutch pedal completely with left foot

2.3.1.2.3 release pressure from accelerator pedal

2.3.1.2.4 move gear stick to next highest gear position

Plan 2.3.1.2.4 - do 1 OR 2 OR 3 OR 4

2.3.1.2.4.1 shift up into 2nd gear

2.3.1.2.4.2 shift up into 3rd gear

2.3.1.2.4.3 shift up into 4th gear

2.3.1.2.4.4 shift up into 5th gear

2.3.1.2.5 raise clutch pedal (smoothly)

2.3.1.2.6 start applying pressure on accelerator

2.3.1.2.7 place left hand back on steering wheel

2.3.1.2.8 raise clutch pedal fully

2.3.1.2.9 depress accelerator an amount proportional to the desired speed or rate of acceleration

2.3.1.2.10 remove foot from clutch pedal

2.3.2 increase current speed

Plan 2.3.2 - do 1. IF greater rate of acceleration required AND IF engine speed >4500RPM THEN 2 THEN repeat step 1. IF rate of speed increase still inadequate AND engine speed >4500RPM THEN repeat 2.

2.3.2.1 depress accelerator pedal an amount proportional to desired rate of acceleration

2.3.2.2 change down a gear

Plan 2.3.2.2 - do 1,2,3 in order. do 4 OR 5 as required. IF strong resistance to gear selection felt at the gear stick THEN 6. WHILE 10 do 7 AND 8 THEN 9

2.3.2.2.1 place left hand on gear stick

2.3.2.2.2 depress clutch pedal completely with left foot

2.3.2.2.3 maintain (a little) pressure on accelerator pedal

2.3.2.2.4 move the gear lever to next lowest gear position

2.3.2.2.5 initiate block change into desired gear

Plan 2.3.2.2.5 - WHILE 1, do 2 OR 3 OR 5 as required. IF speed >5mph THEN 4 OR 6 OR 7 as required

   2.3.2.2.5.1 allow time for gearbox synchronizers to operate

   Plan 2.3.2.2.5.1 - IF gear lever just about to enter desired selector gate THEN 1. IF resistance to gear selection subsides (rapidly) THEN 2 ELSE 3

      2.3.2.2.5.1.1 maintain (gentle) pressure on gear lever at entrance to required gear selector gate

      2.3.2.2.5.1.2 engage gear lever in appropriate selector gate (decisively)

   2.3.2.2.5.2 move gear lever from 5th gear position to 3rd gear position

   2.3.2.2.5.3 move gear lever from 5th gear position to 2nd gear position

   2.3.2.2.5.4 move gear lever from 5th gear position to 1st gear position

   2.3.2.2.5.5 move gear lever from 4th gear position to 2nd gear position

   2.3.2.2.5.6 move gear lever from 4th gear position to 1st gear position

   2.3.2.2.5.7 move gear lever from 3rd gear position to 1st gear position

2.3.2.2.6 initiate double-declutch manoeuvre

Plan 2.3.2.2.6 - do in order (briskly and smoothly)

   2.3.2.2.6.1 move gear stick into neutral position (briskly)

   2.3.2.2.6.2 raise clutch (briefly)

   2.3.2.2.6.3 blip throttle (briskly)

   Plan 2.3.2.2.6.3 - WHILE 2 do 1

      2.3.2.2.6.3.1 depress and release accelerator pedal (momentarily)

      2.3.2.2.6.3.2 keep left foot in contact with pedal during procedure

2.3.2.2.6.4 depress clutch (rapidly)

2.3.2.2.6.5 select desired gear (briskly)

2.3.2.2.7 let clutch pedal come up (smoothly)

2.3.2.2.8 match engine speed with road speed

Plan 2.3.2.2.8 - WHILE 1 do 2

2.3.2.2.8.1 press down a little more on the accelerator pedal

2.3.2.2.8.2 raise engine revs to a level congruent with selected gear ratio

2.3.2.2.9 depress accelerator pedal an amount proportional to the desired rate of acceleration

2.3.2.2.10 put left hand back on the steering wheel

2.3.3 maintain speed

Plan 2.3.3 - do 1. IF increasing pressure on accelerator pedal insufficient to maintain road speed AND IF engine speed >4500RPM THEN 2,3. IF speed maintenance still inadequate THEN repeat plan until desired rate of acceleration is achieved OR vehicles maximum accelerative abilities are reached

2.3.3.1 increase pressure on accelerator to maintain constant speed

2.3.3.2 change down a gear << GO TO subroutine 2.3.2.2 'change down a gear' >>

2.3.3.3 apply sufficient pressure on accelerator to maintain desired speed

2.3.4 control vehicle speed at extremely low speeds (parking, crawling in traffic etc.)

Plan 2.3.4 - do 1,2,3 IF engine note changes THEN 4

2.3.4.1 << GO TO subroutine 2.1.1.1 'set controls ready for pulling away' >>

2.3.4.2 depress accelerator (slightly)

2.3.4.3 raise clutch (slowly)

2.3.4.4 slip clutch manoeuvre

Plan 2.3.4.4 - IF speed <5mph THEN 4. IF speed >5mph THEN 3 AND 1. IF engine racing THEN 4 AND 1 if necessary. IF engine faltering THEN 2 AND 3

2.3.4.4.1 raise clutch (slightly)

2.3.4.4.2 depress clutch (slightly)

2.3.4.4.3 depress accelerator (slightly)

2.3.4.4.4 ease off accelerator (slightly)

2.3.5 avoid abrupt changes in accelerator pressure

2.4 decrease vehicle speed

Plan 2.4 - IF easing back OR gently regulating speed OR maintaining speed by slowing slightly THEN 1. IF decreasing speed to new target speed THEN 1,2. IF coming to complete halt THEN 1,2,3 in order. IF waiting after coming to complete halt THEN 4

2.4.1 Initial slowing

Plan 2.4.1 - do 1 THEN 2

2.4.1.1 << GO TO subroutine 5.1.1.2 'rearward surveillance' >>

2.4.1.2 take right foot off accelerator

2.4.2 decelerate

Plan 2.4.2 - WHILE 1 do 2,3,4,5

2.4.2.1 grip steering wheel with both hands

2.4.2.2 'cover' brake pedal with right foot

Plan 2.4.2.2 - do 1 THEN 2

2.4.2.2.1 place right foot squarely on the brake pedal

2.4.2.2.2 take up initial free movement of the pedal (gently)

2.4.2.3 increase pressure on brake pedal (progressively)

2.4.2.4 monitor rate of deceleration

Plan 2.4.2.4 - IF not slowing quickly enough THEN 1. IF slowing too rapidly THEN 2

2.4.2.4.1 increase pressure on brake pedal (gradually)

2.4.2.4.2 ease off brake pedal (slightly)

2.4.2.5 relax pedal pressure as unwanted road speed is lost

2.4.3 come to a halt

Plan 2.4.3 - WHILE 1 IF speed is >5mph AND/OR engine speed is ~1500RPM THEN 2,3,4. IF vehicle at a complete halt THEN 5

2.4.3.1 retain both hands on steering wheel at quarter to 3 position

2.4.3.2 depress clutch just prior to engine faltering

2.4.3.3 ease off brake pedal

2.4.3.4 release pressure on the brake pedal at instant before stopping

2.4.3.5 reapply brake by an amount sufficient to hold vehicle stationary in position

2.4.4 maintain stop (wait)

Plan 2.4.4 - WHILE 1 THEN 2. IF wait is longer than
                   15 seconds THEN 3

   2.4.4.1 maintain pressure on brake pedal

   2.4.4.2 apply handbrake

   Plan 2.4.4.2 - do in order (briskly)

      2.4.4.2.1 place left hand on handbrake lever

      2.4.4.2.2 depress handbrake knob with thumb of left hand

      2.4.4.2.3 pull lever up from floor (firmly)

      2.4.4.2.4 release handbrake lever knob

      2.4.4.2.5 pull up lever one further ratchet click

      2.4.4.2.6 release foot brake

   2.4.4.3 put in neutral

   Plan 2.4.4.3 - do in order

      2.4.4.3.1 place left hand on gear stick

      2.4.4.3.2 depress clutch with right foot

      2.4.4.3.3 place gear stick in neutral position

      2.4.4.3.4 raise clutch pedal fully (slowly)

2.5 undertake directional control

Plan 3.1 - WHILE 1 AND 2 do 3 AND 4

   2.5.1 keep eyes focused well ahead to anticipate steering corrections

   2.5.2 correct errors in vehicle trajectory (gently)

   Plan 2.5.2 - WHILE 1 do 2 AND 3

      2.5.2.1 maintain hand hold on steering wheel at quarter to 3

      2.5.2.2 make small steering corrections (gradually)

      2.5.2.3 decrease magnitude of corrections as vehicle speed
            increases

   2.5.3 maintain correct position in lane

   Plan 2.5.3 - do 1 ELSE IF oncoming traffic AND/OR road
             conditions dictate THEN 2

      2.5.3.1 generally maintain vehicle in centre of traffic lane

      2.5.3.2 bias vehicle towards left of traffic lane

   2.5.4 deal with road side hazards (e.g. parked cars, kerbs etc.)

   Plan 2.5.4 - do 1 ELSE IF there are no oncoming cars
             THEN 2 ELSE 3

      2.5.4.1 assess ability to pass between obstructions

      2.5.4.2 move to position in which to gain maximum clearance

      2.5.4.3 move tight to left hand obstructions

2.6 negotiate bends

Plan 2.6 - do in order

  2.6.1 undertake information phase

  Plan 2.6.1 - alternate as necessary between (1 AND 2 AND 5 AND 6) OR (do 3 AND 4)

    2.6.1.1 observe for road signs warning of curves

    2.6.1.2 assess environment ahead for indication of a curve

    Plan 2.6.1.2 - do 1 AND 2

      2.6.1.2.1 look into the distance at features such as hedgerows and streetlights to gain advanced warning of the course of the road and the severity of bends

      2.6.1.2.2 look into the distance for natural barriers such as rivers or large motorways, indicating a future bend or change of course

    2.6.1.3 assess traffic in front of vehicle << GO TO subroutine 5.1 'surveillance' >>

    2.6.1.4 assess traffic behind vehicle << GO TO subroutine 5.1 'surveillance' >>

    2.6.1.5 assess road surface conditions

    Plan 2.6.1.5 - do 1 AND 2

      2.6.1.5.1 observe for camber/super elevation

      2.6.1.5.2 observe for factors that may influence available tyre friction

    2.6.1.6 see through bend

    Plan 2.6.1.6 - do 1 AND 2 (IF limit/vanishing point of bend appears to get closer as bend is approached THEN bend is sharp. IF limit/vanishing point of turn appears to get further away as bend is approached THEN bend opens out)

      2.6.1.6.1 observation of curve features

      2.6.1.6.2 observation of limit/vanishing point

  2.6.2 undertake position phase

  Plan 2.6.2 - do 1 AND 2

    2.6.2.1 position car appropriately for best view through turn

    Plan 2.6.2.1 - IF right hand bend THEN 1. IF left hand bend THEN 2

      2.6.2.1.1 position car appropriately for right-hand bends

      Plan 2.6.2.1.1 - do 1 AND 2

2.6.2.1.1.1 position car towards the left of your road space

2.6.2.1.1.2 avoid coming into conflict with nearside hazards

Plan 2.6.2.1.1.2 - when required do 1 AND/OR 2 AND/OR 3 AND/OR 4 AND/OR 5

2.6.2.1.1.2.1 give safe clearance for parked vehicles

2.6.2.1.1.2.2 give safe clearance for pedestrians

2.6.2.1.1.2.3 give safe clearance for cyclists

2.6.2.1.1.2.4 check for blind junctions or exits

2.6.2.1.1.2.5 avoid adverse cambers or poor nearside road surface conditions

2.6.2.1.2 position car appropriately for left hand turns

Plan 2.6.2.1.2 - do 1 WHILE 2 AND 3

2.6.2.1.2.1 position car towards the centre line of lane

2.6.2.1.2.2 avoid coming into conflict with oncoming traffic

2.6.2.1.2.3 avoid misleading other drivers as to intentions whilst repositioning vehicle

2.6.2.2 (generally) avoid crossing central lane divide

2.6.3 undertake speed phase

Plan 2.6.3 - do 1. IF current speed > target speed THEN 2

2.6.3.1 assess ability to stop in distance between car and limit/vanishing point

2.6.2.2 decelerate to target speed << GO TO subroutine 2.4 'decrease speed' >>

2.6.4 undertake gear phase

Plan 2.6.4 - do 1 ELSE 2

2.6.4.1 select gear that is appropriate for target speed << GO TO subroutine 2.3.2.2 'change down a gear' >>

2.6.4.2 remain in current gear

2.6.5 undertake acceleration phase

Plan 2.6.5 - WHILE 1 AND 2 AND 3 do 4 AND 5 OR 6 ELSE 7

2.6.5.1 turn steering wheel an amount proportional to the desired rate of turn << GO TO subroutine 2.2 'steering' >>

2.6.5.2 look well ahead of turning path

2.6.5.3 avoid harsh control inputs mid turn

2.6.5.4 use accelerator during turn

Plan 2.6.5.4 - IF no additional hazards AND limit/vanishing point begins to move away AND steering is beginning to be straightened THEN 1. IF steering is continuing to be straightened AND speed limit is not yet attained OR original pre-corner speed is not yet attained OR desired speed is not yet attained AND/OR no other considerations restrict speed THEN 2 to 'catch the limit/vanishing point' of the turn

2.6.5.4.1 begin accelerating (gently)

2.6.5.4.2 increase acceleration

2.6.5.5 reduce the curvature of right hand bends

Plan 2.6.5.5 - WHILE 4 IF clear view across bend is available AND there is no oncoming traffic THEN 1 AND 2. IF apex of bend has been passed THEN 3

2.6.5.5.1 take a curving path towards the centre of the road (gradually)

2.6.5.5.2 bring vehicle closest to centre line at apex of bend

2.6.5.5.3 bring vehicle back towards normal road position (gently)

2.6.5.5.4 present no additional risk to other road users

2.6.5.6 reduce the curvature of left-hand bends

Plan 2.6.5.6 - WHILE 5 do 1 IF view ahead is clear THEN 2 THEN 3 THEN 4

2.6.5.6.1 keep car towards centre line

2.6.5.6.2 take a curving path towards nearside of road (gradually)

2.6.5.6.3 bring vehicle closest to nearside at apex of bend

2.6.5.6.4 bring vehicle back towards normal road position (gently)

2.6.5.6.5 present no additional risk to other road users

2.6.5.7 steer so as to maintain normal position in lane

2.7 negotiate gradients

Plan 2.7 - do 1 OR 2 as required

2.7.1 negotiate uphill gradient

Plan 2.7.1 - do in order

2.7.1.1 approach uphill gradient

Plan 2.7.1.1 - do 1 AND IF sufficient view ahead THEN 2.
IF uphill gradient >1:20 AND engine speed
>3500RPM THEN 3 AND 4 as necessary to
maintain speed. IF other vehicles on hill are
slowing THEN 5

2.7.1.1.1 observe road signs indicating severity of gradient

2.7.1.1.2 observe other vehicles negotiating hill

2.7.1.1.3 begin accelerating

2.7.1.1.4 << GO TO subroutine 2.3.2.2 'change down a gear' >>

2.7.1.1.5 maintain generous distance when following other
vehicles

2.7.1.2 climb uphill gradient

Plan 2.7.1.2 - WHILE 1 AND 2 do 3 AND 4 as required

2.7.1.2.1 << GO TO subroutine 2.3.3 'maintain speed' >>

2.7.1.2.2 maintain engine revs between 3500-5500RPM
(select lowest engine speed within this range
that enables the car to climb without the engine
labouring)

2.7.1.2.3 monitor engine parameters on long inclines

Plan 2.7.1.2.3 - do 1 AND 2

2.7.1.2.3.1 monitor engine temperature gauge

2.7.1.2.3.2 monitor engine sounds

2.7.1.2.4 respond to monitored engine parameters

Plan 2.7.1.2.4 - IF engine temperature nearing red zone
THEN 1 AND 2 will assist engine cooling.
IF temperature gauge needle enters red zone
THEN 3 AND 4. IF engine labouring AND/OR
emitting otherwise abnormal sounds or responses
THEN 2. IF engine still emitting abnormal
sounds THEN 3 AND 4.

2.7.1.2.4.1 switch on interior heater

2.7.1.2.4.2 reduce engine load

Plan 2.7.1.2.4.2 - do 1 AND/OR 2

2.7.1.2.4.2.1 reduce throttle opening

2.7.1.2.4.2.2 << GO TO subroutine 2.3.2.2 'change
down a gear' >>

2.7.1.2.4.3 pull off road

2.7.1.2.4.4 seek assistance

2.7.1.3 approach crest of uphill gradient

Plan 2.7.1.3 - IF uphill gradient steep enough to cause sig-
nificant over acceleration upon cresting summit
THEN 1. IF gradient caused reduction in speed
THEN 2 to target speed ELSE 3. IF road narrow
THEN 4 AND 5 if necessary

  2.7.1.3.1 decelerate slightly

  2.7.1.3.2 << GO TO subroutine 2.3.2 'increase current
  speed' >>

  2.7.1.3.3 << GO TO subroutine 2.3.3 'maintain current
  speed' >>

  2.7.1.3.4 keep too far left of road

  2.7.1.3.5 sound horn to alert oncoming vehicles

2.7.2 negotiate downhill gradient

Plan 2.7.2 - do in order

  2.7.2.1 approach downhill gradient

  Plan 2.7.2.1 - do 1 AND 2 IF signs OR road characteristics
  dictate THEN 3 AND/OR 4 in anticipation

    2.7.2.1.1 look for signs indicating length/gradient of
    downhill

    2.7.2.1.2 check characteristics of downhill gradient

    Plan 2.7.2.1.2 - do 1 AND 2

      2.7.2.1.2.1 check length of downhill gradient

      2.7.2.1.2.2 check gradient of downhill

    2.7.2.1.3 << GO TO subroutine 2.3.2.2 'change down a
    gear' >>

    2.7.2.1.4 << GO TO subroutine 2.4 'decrease speed' >>

  2.7.2.2 descend downhill gradient

  Plan 2.7.2.2 - WHILE 1 AND 2 IF reduction in speed is
  required THEN 3. IF speed reduction inadequate
  THEN 4. IF hill very steep, OR if not completed
  on approach to descent THEN 5

    2.7.2.2.1 maintain constant speed

    2.7.2.2.2 give right of way to climbing vehicles as necessary

    2.7.2.2.3 reduce accelerator pressure

    2.7.2.2.4 use brakes

    Plan 2.7.2.2.4 - IF descent is relatively short THEN 1 as
    required. IF descent is long AND/OR very steep

THEN 2 THEN steps 3,4,5 in order AND repeat for duration of very steep descent

2.7.2.2.4.1 << GO TO subroutine 2.4 'decrease speed' >>

2.7.2.2.4.2 release pressure on accelerator

2.7.2.2.4.3 apply brake pedal with left foot an amount proportional to the desired rate of deceleration for 6 seconds (progressively)

2.7.2.2.4.4 completely release pressure on brake pedal for 2 seconds (smoothly)

2.7.2.2.4.5 reapply brakes an amount proportional to the desired rate of deceleration (progressively)

2.7.2.2.5 << GO TO subroutine 2.3.2.2 'change down a gear' >>

2.7.2.3 approach bottom of descent

Plan 2.7.2.3 - IF speed of descent < desired target speed THEN 1. IF speed of descent > desired target speed THEN 2 ELSE 3

2.7.2.3.1 << GO TO subroutine 2.3.2 'increase current speed' >>

2.7.2.3.2 << GO TO subroutine 2.4 'decrease speed' >>

2.7.2.3.3 << GO TO subroutine 2.3.3 'maintain speed' >>

2.8 reverse the vehicle

Plan 2.8 - do in order

2.8.1 prepare to back up

Plan 2.8.1 - do in order

2.8.1.1 ensure car is fully stopped

2.8.1.2 scan area for suitability and obstructions

Plan 2.8.1.2 - do in any order

2.8.1.2.1 glance in rear-view mirror

2.8.1.2.2 look over right shoulder

2.8.1.2.3 look over left shoulder

2.8.1.2.4 check side mirrors

2.8.1.3 shift into reverse

Plan 2.8.1.3 - do in order

2.8.1.3.1 place hand on gear stick

2.8.1.3.2 depress clutch

2.8.1.3.3 place gear stick in reverse gear position (following manufacturer's procedures)

2.8.1.4 release handbrake << GO TO subroutine 2.1.1.2.8
    'release handbrake' >>

2.8.1.5 assume correct body position

Plan 2.8.1.5 - IF reversing to the right THEN 3 ELSE 1 AND 2

   2.8.1.5.1 turn upper body to face left of vehicle

   2.8.1.5.2 turn head to look out of rear window

   2.8.1.5.3 turn head over right shoulder

2.8.2 back up

Plan 2.8.2 - do 1 AND 2 AND 3 as required

   2.8.2.1 << GO TO subroutine 2.3.4 'extremely low speeds' >>

   2.8.2.2 steering

   Plan 2.8.2.2 - WHILE 1 do 2 THEN 3

      2.8.2.2.1 avoid quick steering corrections

      2.8.2.2.2 maintain grip on steering wheel

      2.8.2.2.3 turning

      Plan 2.8.2.2.3 - WHILE 3 AND 4 do 1 AND 2

         2.8.2.2.3.1 turn top of steering wheel to the side of the
             rear of the car that is to move

         2.8.2.2.3.2 << GO TO subroutine 2.2.6 'reversing hold' >>

         2.8.2.2.3.3 proceed slowly

         2.8.2.2.3.4 observe the front of the vehicle (frequently)

   2.8.2.3 use rear-view mirrors where view can be enhanced

2.8.3 complete reversing manoeuvre

Plan 2.8.3 - WHILE 1, do 2,3,4,5. IF vehicle stationary THEN 6

   2.8.3.1 allow greater stopping distances in reverse

   2.8.3.2 depress clutch

   2.8.3.3 release accelerator immediately

   2.8.3.4 depress brake

   2.8.3.5 shift from reverse to neutral

   2.8.3.6 apply handbrake << GO TO subroutine 2.4.4.2 'apply
       handbrake' >>

3: PERFORM OPERATIONAL DRIVING TASKS

Plan 3 - do 1 OR 2 OR 3 as required. do 4 THEN 5 THEN 6
THEN 7

3.1 emerge into traffic from side of road

Plan 3.1 - do 1,2,3

   3.1.1 observe traffic situation

Plan 3.1.1 - do WHILE 1, do 2,3,4,5,6

    3.1.1.1 give way to rear-approaching traffic

    3.1.1.2 look for a suitable gap in traffic

    3.1.1.3 note vehicle that car will enter behind

    3.1.1.4 traffic surveillance << GO TO subroutine 5.1 'surveillance' >>

    3.1.1.5 check blind spots by glancing over shoulder

    3.1.1.6 activate appropriate indicator as vehicle passes

3.1.2 enter traffic lane

Plan 3.1.2 - do 1 THEN 2, IF setting off from right side of the road THEN 3

    3.1.2.1 << GO TO subroutine 2.1 'pulling away from standstill' >>

    3.1.2.2 accelerate (smoothly/briskly) into traffic lane << GO TO subroutine 2.3.1 'accelerate from rest' >>

    3.1.2.3 turn steering wheel enough to cross road at sharp angle

3.1.3 establish vehicle in new lane

Plan 3.1.3 - do 1 AND 2 AND 3 as required

    3.1.3.1 straighten steering wheel

    3.1.3.2 check to be sure that indicator has cancelled

    3.1.3.3 accelerate quickly to attain speed of traffic flow << GO TO subroutine 2.3.2 'increase current speed' >>

3.2 follow other vehicles

Plan 3.2 - do 1 AND 3 THEN 2 as required

    3.2.1 maintain a safe headway separation

    Plan 4.2.1 - do 1 AND 2 to estimate safe following distance. IF travelling on fast roads THEN 3 THEN 4 IF car passes same landmark >2 seconds THEN 5 and repeat at step 3 as necessary. IF following oversized vehicles/public service vehicles (or other vehicles that stop frequently)/two wheeled vehicles/vehicles driving erratically AND/OR in poor weather/ visibility/at night/where traffic intersects THEN 5

    3.2.1.1 use knowledge of safe braking distances as provided in the Highway Code

    3.2.1.2 embody thinking and actual braking distance in any estimates

3.2.1.3 note when the lead car passes a convenient roadside
          landmark

3.2.1.4 count two seconds

3.2.1.5 increase separation distance (to beyond 2 seconds
          headway)

3.2.2 adjust speed to changes in speed of lead vehicle

Plan 3.2.2 - do 1, IF (rapid) closure of headway detected AND/
          OR lead vehicle otherwise indicates a reduction in
          speed THEN 2

3.2.2.1 check for indications of reduced speed of lead vehicle

Plan 3.4.2.1 - do 1 AND 2 AND 3 AND 4 AND 5 AND 6
          (5 & 6 provide tentative supplementary queues to
          speed reduction of the lead vehicle)

3.2.2.1.1 gauge closure of headway/relative speeds

3.2.2.1.2 observe lead vehicles indicators

3.2.2.1.3 observe lead vehicles brake lights

3.2.2.1.4 observe driver of lead vehicle for hand signals
            indicating reduction in speed

3.2.2.1.5 look for the front of the lead vehicle dipping in
            response to brake application

3.2.2.1.6 look for visible indications from lead vehicle's
            exhaust (puffs of smoke etc.) that might serve as
            an indication the lead vehicle's throttle has just
            been closed abruptly

3.2.2.2 << GO TO subroutine 2.4 'decrease speed' >>

3.2.3 observe specific traffic conditions to anticipate changes in
          lead vehicle velocity

Plan 3.2.3 - do 1 AND 2 AND 3. IF traffic conditions dictate
          reduction in speed of lead vehicle THEN 4 AND 5

3.2.3.1 check vehicles in front of lead vehicle

3.2.3.2 monitor lead vehicle deceleration due to junctions

3.2.3.3 monitor lead vehicle deceleration due to road layout

3.2.3.4 prepare to slow

3.2.3.5 << GO TO subroutine 2.4 'decrease speed' >>

3.3 overtake other moving vehicles

Plan 3.3 - WHILE 1 do 2 THEN 3 THEN 4 THEN 5 THEN 6.
IF overtaking in a stream of vehicles THEN 7 THEN 6

3.3.1 generally avoid overtaking on the left

3.3.2 << GO TO subroutine 3.2 'following' >>

3.3.3 assess the road and traffic conditions for an opportunity to overtake safely

Plan 3.3.3 - WHILE 1 do 2 AND 3 AND 4 AND 5 THEN 6 AND 7. IF any of these parameters render the completion of the passing manoeuvre doubtful/unsafe THEN 8

  3.3.3.1 observe oncoming traffic

  Plan 3.3.3.1 - do 1 AND 2 AND 3 AND 4 AND 5

    3.3.3.1.1 judge distance from first oncoming vehicle

    3.3.3.1.2 judge lead vehicles relative speed

    3.3.3.1.3 judge available passing time

    3.3.3.1.4 determine whether pass can be completed with available passing distance

    3.3.3.1.5 observe for 'lurkers' stuck close behind slower oncoming vehicles

  3.3.3.2 assess road conditions

  Plan 3.3.3.2 - do 1 AND 2 AND 3 AND 4

    3.3.3.2.1 anticipate vehicles pulling out of nearside junctions, slip-roads, laybys, driveways, paths and tracks, farm entrances etc.

    3.3.3.2.2 anticipate vehicles pulling out of offside junctions, slip roads, laybys, driveways, paths and tracks, farm entrances etc.

    3.3.3.2.3 observe road features such as bridges, bends, hill crests etc. that obscure forward view

    3.3.3.2.4 observe road surface for ruts, holes, adverse cambers/super elevations, or surface water

  3.3.3.3 check lead vehicle is not about to change its speed

  Plan 3.3.3.3 - do 1 AND 2 AND 3

    3.3.3.3.1 observe any clearing traffic in front of lead vehicle

    3.3.3.3.2 observe road features that may cause lead vehicle to accelerate

    3.3.3.3.3 gauge future driving actions based on lead driver's previous actions

  3.3.3.4 assess gap ahead of lead vehicle

  3.3.3.5 anticipate course of lead vehicle

Plan 3.3.3.5 - do 1 AND 2 AND 3

    3.3.3.5.1 check lead vehicle is not indicating or about to turn

    3.3.3.5.2 check lead vehicle is not passing cyclists/animals etc.

    3.3.3.5.3 gauge future driving actions based on lead driver's previous actions

    3.3.3.6 use stored mental representations of speed and performance of own vehicle

    3.3.3.7 assess availability of safety margin should manoeuvre be aborted

    3.3.3.8 decide not to overtake

3.3.4 adopt the overtaking position

Plan 3.3.4 - do 1 THEN 2 AND 3 THEN 4

    3.3.4.1 undertake information phase

Plan 3.3.4.1 - WHILE 1 AND 2 IF an overtaking opportunity begins to develop THEN 3 AND 4

    3.3.4.1.1 continue observing road ahead << GO TO subroutine 5.1 'surveillance' >>

    3.3.4.1.2 continue observing road behind (periodically) << GO TO subroutine 5.1 'surveillance' >>

    3.3.4.1.3 plan overtaking move

    3.3.4.1.4 consider the need to indicate

    3.3.4.2 undertake position phase

Plan 3.3.4.2 - WHILE 1 IF following normal sized vehicles AND view ahead is clear THEN 2. IF following large vehicles THEN 3 AND 4 if required to gain clear view ahead

    3.3.4.2.1 move into overtaking position

Plan 3.3.4.2.1 - WHILE 1 do 2 AND in order to gain best forward view do 3 OR 4 as required. IF overtaking manoeuvre involves a right-hand bend THEN 5 to gain clear view through turn THEN 6. IF overtaking manoeuvre involves a left-hand bend THEN 7 (to gain clear view through turn) THEN 8

    3.3.4.2.1.1 maintain adequate view of road ahead

    3.3.4.2.1.2 reduce headway from normal following situations

3.3.4.2.1.3 position car to left of available lane space

3.3.4.2.1.4 position car to right of available lane space

3.3.4.2.1.5 position car to extreme nearside of available lane space

3.3.4.2.1.6 move up to lead vehicle as it approaches apex of bend

3.3.4.2.1.7 maintain a position where a clear view along the nearside of the lead vehicle is possible

3.3.4.2.1.8 move out to offside of available lane space

3.3.4.2.2 continue reducing headway consistent with hazards

3.3.4.2.3 accept larger headway consistent with forward view of road

3.3.4.2.4 move to the left and right (slightly) to gain view down both sides of the obstructing vehicle

3.3.4.3 undertake speed phase

Plan 3.3.4.3 - WHILE 1 do 2

3.3.4.3.1 avoid 'tailgating' and intimidating lead vehicle

3.3.4.3.2 adjust speed to that of vehicle in front (smoothly)

Plan 3.3.4.3.2 - IF lead vehicle speed > car speed THEN 1. IF lead vehicle speed < car speed THEN 2

3.3.4.3.2.1 << GO TO subroutine 2.3.2 'increase current speed' >>

3.3.4.3.2.2 << GO TO subroutine 2.4 'decrease speed' >>

3.3.4.4 undertake gear phase

Plan 3.3.4.4 - do 1 ELSE 2

3.3.4.4.1 select the most responsive gear for completing the overtaking manoeuvre

Plan 3.3.4.4.1 - do 1 THEN 2

3.3.4.4.1.1 assess speed and performance of own vehicle in relation to gear choice and road speed

3.3.4.4.1.2 << GO TO subroutine 2.3.2.2 'change down a gear' >>

3.3.4.4.2 remain in current gear

3.3.5 perform overtaking manoeuvre

Plan 3.3.5 - do 1 THEN 2 THEN 3

3.3.5.1 undertake information phase

Plan 3.3.5.1 - do 1 AND 2 AND 3 AND 4 THEN 5

3.3.5.1.1 recheck that there is adequate vision along particular stretch of road

3.3.5.1.2 recheck that there is a gap ahead in which to safely return to after completing the overtake

3.3.5.1.3 recheck the speed of any approaching vehicles

3.3.5.1.4 recheck relative speed of own vehicle and the vehicle(s) ahead that are to be overtaken

3.3.5.1.5 recheck what is happening behind

Plan 3.3.5.1.5 - do 1 AND 2. do 3 IF view behind is enhanced. IF immediately prior to entering position phase THEN 4

3.3.5.1.5.1 glance in rear-view mirror

3.3.5.1.5.2 glance in offside side mirror

3.3.5.1.5.3 glance in nearside side mirror

3.3.5.1.5.4 check blind spot by glancing (quickly) over right shoulder

3.3.5.2 undertake position phase

Plan 3.3.5.2 - WHILE 1 do 2 THEN 3 THEN 4. IF results of final safety checks render the completion of the intended manoeuvre doubtful THEN 5 AND 6

3.3.5.2.1 maintain current speed

3.3.5.2.2 use appropriate indicator

3.3.5.2.3 move vehicle completely into offside position

3.3.5.2.4 assess any new information for safety of intended manoeuvre (quickly) << GO TO subroutine 5.1 'surveillance' >>

3.3.5.2.5 move back into left lane (smoothly)

3.3.5.2.6 abandon manoeuvre

3.3.5.3 undertake speed phase

Plan 3.3.5.3 - WHILE 1 AND 2 do 3

3.3.5.3.1 maintain both hands on the steering wheel

3.3.5.3.2 generally avoid changing gear during overtaking manoeuvre

3.3.5.3.3 accelerate past vehicles to be overtaken decisively (briskly)

Plan 3.3.5.3.3 - WHILE 1 do 2

3.3.5.3.3.1 use full range of engine performance as required

Plan 3.3.5.3.3.1 - WHILE 1 AND 2 do 3 AND 4 if
required to complete the manoeuvre safely

3.3.5.3.3.1.1 maintain engine speed above at least
3500RPM for duration of manoeuvre

3.3.5.3.3.1.2 do not change up a gear too early

3.3.5.3.3.1.3 allow engine revs to rise to 6500RPM

3.3.5.3.3.1.4 use full throttle

3.3.5.3.3.2 << GO TO subroutine 2.3.2 'increase current
speed' >>

3.3.6 complete overtaking manoeuvre

Plan 3.3.6 - WHILE in overtaking position do 1. IF further
overtaking not undertaken THEN 2 THEN 3

3.3.6.1 consider opportunity for further overtaking

Plan 3.3.6.1 - do 1. IF further overtaking can be completed
safely AND further overtaking is necessary to
assume desired speed post-overtaking THEN 2
OR 3 if situation dictates

3.3.6.1.1 << GO TO subroutine 3.3.4.1 'undertake information phase' >>

3.3.6.1.2 << GO TO subroutine 2.3.3 'maintain speed' >>

3.3.6.1.3 << GO TO subroutine 2.3.2 'increase current
speed' >>

3.3.6.2 enter gap identified during earlier information phases

Plan 3.3.6.2 - WHILE gap is approached do 1 AND do 2 as
required THEN 3 THEN 4 THEN 5 THEN 6

3.3.6.2.1 assess relative speeds of other traffic

3.3.6.2.2 adjust vehicle speed

Plan 3.3.6.2.2 - do 1 OR 2 as required to merge seamlessly
with nearside traffic

3.3.6.2.2.1 << GO TO subroutine 2.4 'decrease speed' >>

3.3.6.2.2.2 << GO TO subroutine 2.3.3 'maintain speed' >>

3.3.6.2.3 << GO TO subroutine 5.1 'surveillance' >>

3.3.6.2.4 use appropriate indicator

3.3.6.2.5 move progressively into left-hand lane

3.3.6.2.6 assume normal road position

3.3.6.3 resume desired speed in normal driving lane

Plan 3.3.6.3 - IF overtaking manoeuvre particularly energetic
THEN 1 ELSE 2 WHILE 3

3.3.6.3.1 << GO TO subroutine 2.4 'decrease speed' >>

3.3.6.3.2 << GO TO subroutine 2.3.3 'maintain speed' >>

3.3.6.3.3 observe posted speed limit for road

3.3.7 overtake in a stream of vehicles

Plan 3.3.7 - do 1 AND 2 AND 3

3.3.7.1 take into account possible actions of lead drivers in overtaking stream

3.3.7.2 take into account possible actions of following drivers in overtaking stream

3.3.7.3 ensure there is a suitable 'escape route' at every stage of overtaking in a stream of vehicles

Plan 3.3.7.3 - WHILE 5 IF no oncoming traffic OR closing speed of oncoming traffic not a threat to the completion of the manoeuvre THEN 1. IF oncoming traffic closing rapidly OR any other traffic or road conditions diminish safety THEN 2. IF lead vehicles in stream making progress past nearside traffic THEN 3 THEN 4. repeat steps 3 AND 4 for duration of passing nearside traffic

3.3.7.3.1 maintain offside road position

3.3.7.3.2 be prepared to head into adjacent gap in nearside traffic

3.3.7.3.3 hold offside road position adjacent to gap in nearside traffic

3.3.7.3.4 proceed to next adjacent observed gap in nearside traffic

3.3.7.3.5 avoid cutting in on nearside traffic

3.4 approach junctions (crossings/intersections etc.)

Plan 3.4 - IF junction comes into view THEN 1. IF junction > 50 meters ahead OR junction information signs in clear view AND/OR junction road markings present THEN 2. IF junction is a give way OR other road situations dictate AND/OR other traffic situations dictate THEN 3

3.4.1 undertake information phase

Plan 3.4.1 - do 1 THEN 2 AND 3 THEN 4

3.4.1.1 assess conditions behind using rear-view mirrors

<< GO TO subroutine 5.1 'surveillance' >>

3.4.1.2 assess road features of junction

Plan 3.4.1.2 - do 1,2,3,4,5,6,7 as information becomes available

    3.4.1.2.1 assess configuration of junction

    3.4.1.2.2 assess number of roads involved in junction

    3.4.1.2.3 assess width of roads involved in junction

    3.4.1.2.4 assess condition of roads involved in junction

    3.4.1.2.5 assess possible gradients of roads involved in junction

    3.4.1.2.6 assess the comparative importance of any side roads being passed in relation to the road being travelled on currently

    3.4.1.2.7 consider visibility into side roads being passed

3.4.1.3 assess traffic features of junction

Plan 3.4.1.3 - do 1 AND 2 AND 3

    3.4.1.3.1 consider the presence of other road users using junction

    3.4.1.3.2 assess road signs on approach to junction

    3.4.1.3.3 assess road markings on approach to junction

3.4.1.4 plan strategy for dealing with junction

Plan 3.4.1.4 - do 1,2,3,4,5,6,7 in order that is appropriate for the situation

    3.4.1.4.1 consider need to alter vehicles position on road

    Plan 3.4.1.4.1 - do 1 AND/OR 2 AND/OR 3 AND/OR 4

        3.4.1.4.1.1 assess signs indicating route guidance

        3.4.1.4.1.2 assess signs indicating lane selection

        3.4.1.4.1.3 assess road markings advising route selection

        3.4.1.4.1.4 assess road markings advising lane selection

    3.4.1.4.2 consider need to respond to relevant signals or traffic controls

    3.4.1.4.3 consider need to respond to stop signs

    3.4.1.4.4 consider need to respond to give way signs or give way lines

    3.4.1.4.5 consider what road appears to be the busiest in the junction

    3.4.1.4.6 consider whether the need to stop is obvious

    3.4.1.4.7 consider whether to continue forward until need to stop manifests itself

3.4.2 apply mirror-signal-manoeuvre routine

Plan 3.4.2 - do 1 THEN 2 THEN 3 (based on 4.4.1.4 above)
IF conditions dictate THEN 4 AND 5 OR 4
THEN 5 OR 5 THEN 4. IF vehicle now in junc-
tion THEN 6

3.4.2.1 assess rearward traffic conditions using rear view mirrors << GO TO 5.1 'surveillance' >>

3.4.2.2 check blind spots

Plan 3.4.2.2 - IF manoeuvre to the right is planned THEN 1.
IF manoeuvre to the left is planned THEN 2

3.4.2.2.1 glance over right shoulder

3.4.2.2.2 glance over left shoulder

3.4.2.3 activate left or right indicator

3.4.2.4 undertake position phase

Plan 3.4.2.4 - (based on 4.4.1.4 above, select correct position
for desired exit/route through and out of junction)
do 1 THEN 2

3.4.2.4.1 begin manoeuvre into new position/lane (gently)

3.4.2.4.2 move decisively into new road position/lane

Plan 3.4.2.4.2 - IF driving on single lane road AND right-
hand turn is desired THEN 1. IF driving on
single lane road AND desired course is straight
ahead THEN 2. IF driving on single lane road
AND left turn is desired THEN 3. IF desired
junction turn requires a change of lane THEN 4.
IF approaching a roundabout THEN 5

3.4.2.4.2.1 position vehicle towards centre of road

3.4.2.4.2.2 position vehicle in centre of current lane

3.4.2.4.2.3 position vehicle towards left of current lane

3.4.2.4.2.4 position vehicle in centre of new lane

3.4.2.4.2.5 select correct lane on approach to roundabouts

Plan 3.4.2.4.2.5 - do 1 ELSE IF taking left or early exit
from roundabout THEN 2. IF going straight
on THEN 2 ELSE 3 when left lane full. IF
roundabout approach is only two lane AND exit
straight ahead is desired THEN 4 ELSE 3. IF
later OR right-hand exits are desired THEN 4

3.4.2.4.2.5.1 be guided by road signs and markings

3.4.2.4.2.5.2 select left lane (unless directed otherwise)

3.4.2.4.2.5.3 select centre lane (unless directed otherwise)

3.4.2.4.2.5.4 select right lane (unless directed otherwise)

3.4.2.5 undertake speed phase

Plan 3.4.2.5 - IF traffic OR road conditions dictate THEN 1 OR 3. IF road OR traffic conditions dictate OR approaching give way junction type THEN 2

    3.4.2.5.1 increase speed << GO TO subroutine 2.3.2 'increase current speed' >>

    3.4.2.5.2 decrease speed << GO TO subroutine 2.4 'decrease current speed' >>

    3.4.2.5.3 maintain existing speed << GO TO subroutine 2.3.3 'maintain speed' >>

3.4.2.6 observe traffic situation

Plan 3.4.2.6 - WHILE 1 do 2 THEN 3 THEN 4

    3.4.2.6.1 scan forward scene << GO TO subroutine 5.1 'surveillance' >>

    3.4.2.6.2 look left

    3.4.2.6.3 look right

    3.4.2.6.4 check blind spot (quickly)

    Plan 3.4.2.6.4 - IF turning left THEN 1 IF turning right THEN 2

        3.4.2.6.4.1 glance left

        3.4.2.6.4.2 glance right

3.4.3 bring vehicle to a stop << GO TO subroutine 2.4.2 'decelerate' >>

3.5 deal with junctions

Plan 3.5 - do 1 OR 2 OR 3 OR 4 as required

3.5.1 deal with slip roads

Plan 3.5.1 - do 1 OR 2

    3.5.1.1 negotiate on-slips

    Plan 3.5.1.1 - do 1 THEN 2 THEN 3. IF no safe opportunity presents itself for entering the main carriageway THEN WHILE 4 do 5 THEN 6 THEN WHILE 7 do 8. IF safe opportunity for joining main carriageway arises THEN 9

        3.5.1.1.1 entering on-slip

Plan 3.5.1.1.1 - WHILE 1 do 2 AND 3 AND 4 AND 5
THEN 6 AND 7

3.5.1.1.1.1 observe posted speed limit

3.5.1.1.1.2 watch for warning signs to slow down or give
way

3.5.1.1.1.3 observe general on-slip/main carriageway
configuration

3.5.1.1.1.4 survey traffic on main carriageway

Plan 3.5.1.1.1.4 - do 1 THEN 2. IF joining traffic lane
from the right OR slip road carries more than one
lane of traffic THEN 3

3.5.1.1.1.4.1 check mirrors

3.5.1.1.1.4.2 look over right shoulder

3.5.1.1.1.4.3 look over left shoulder

3.5.1.1.1.5 evaluate location and speeds of vehicles in
front of car

3.5.1.1.1.6 make initial car speed adjustment based on
on-slip/joining carriageway configuration << GO
TO subroutine 2.3 'speed control' >>

3.5.1.1.1.7 make initial car speed adjustment based on
survey of traffic << GO TO subroutine 2.3 'speed
control' >>

3.5.1.1.2 prepare to join main carriageway

Plan 3.5.1.1.2 - do 1 THEN 2. IF suitable gaps/
opportunities appear THEN 3 WHILE 4 AND 5
AND 6 IF slip road < 2 lanes THEN 7

3.5.1.1.2.1 activate appropriate indicator

3.5.1.1.2.2 look for gaps in main carriageway traffic that
will permit car to merge without interfering with
progress of other vehicles

Plan 3.5.1.1.2.2 - do 1 AND 2

3.5.1.1.2.2.1 assess traffic on main carriageway using
rear-view mirror

3.5.1.1.2.2.2 assess traffic on main carriageway using
appropriate side mirror

3.5.1.1.2.3 adopt speed that will allow car to reach main
roadway coincident with gap

Plan 3.5.1.1.2.3 - do 1 OR 2 ELSE 3

    3.5.1.1.2.3.1 increase speed << GO TO subroutine 2.3.2 'increase current speed' >>

    3.5.1.1.2.3.2 decrease speed << GO TO subroutine 2.4 'decrease speed' >>

    3.5.1.1.2.3.3 maintain existing speed << GO TO subroutine 2.3.3 'maintain speed' >>

3.5.1.1.2.4 periodically recheck main carriageway

Plan 3.5.1.1.2.4 - do 1 AND/OR 2 commensurate with clearest view WHILE 3

    3.5.1.1.2.4.1 use rear-view mirror

    3.5.1.1.2.4.2 use appropriate side mirror

    3.5.1.1.2.4.3 employ occasional rearward glances

3.5.1.1.2.5 recheck position and progress of traffic ahead on on-slip

3.5.1.1.2.6 where practical allow vehicle ahead to enter main carriageway before attempting manoeuvre

3.5.1.1.2.7 avoid overtaking on slip road

3.5.1.1.3 enter main carriageway

Plan 3.5.1.1.3 - do 1, IF at point of entering main carriageway THEN 2 THEN 3 THEN 4 THEN 5 WHILE 6 do 7

    3.5.1.1.3.1 select gap that will permit car to merge onto main roadway without interfering with progress of other vehicles

    3.5.1.1.3.2 final observation checks

Plan 3.5.1.1.3.2 - IF entering carriageway from the left THEN 1. IF entering carriageway from the right THEN 2. IF slip road carries more than one lane of traffic THEN 1 THEN 2

    3.5.1.1.3.2.1 check over left shoulder

    3.5.1.1.3.2.2 check over right shoulder

3.5.1.1.3.3 observe lead vehicle ahead of gap

3.5.1.1.3.4 recheck following vehicle in gap using mirrors

3.5.1.1.3.5 make minor speed adjustments to match speed of lead vehicle behind gap

3.5.1.1.3.6 avoid cutting in on following vehicle behind gap

3.5.1.1.3.7 guide car decisively into adjacent lane of main carriageway (smoothly)

3.5.1.1.4 keep indicators on

3.5.1.1.5 slow car down on slip road << GO TO subroutine 2.4 'decrease speed' >>

3.5.1.1.6 stop vehicle well ahead of end of slip road << GO TO subroutine 2.4.3 'coming to a halt' >>

3.5.1.1.7 glance forward occasionally

3.5.1.1.8 turn head over shoulder

3.5.1.1.9 accelerate hard onto main carriageway << GO TO subroutine 2.1 'pulling away from standstill' >>

3.5.1.2 negotiate off-slips

Plan 3.5.1.2 - do 1 THEN 2

3.5.1.2.1 entering deceleration lane prior to off-slip

Plan 3.5.1.2.1 - WHILE 1 do 2 THEN 3 THEN 4 WHILE 5

3.5.1.2.1.1 guide car smoothly onto off-slip

3.5.1.2.1.2 estimate off-ramp length and curvature

3.5.1.2.1.3 plan extent of deceleration required on off-slip

3.5.1.2.1.4 decrease speed << GO TO subroutine 2.4 'decrease speed' >>

3.5.1.2.1.5 glance at speedometer to ensure appropriate deceleration

3.5.1.2.2 using off-slip

Plan 3.5.1.2.2 - WHILE 1 AND 2 AND 3 do 4 AND 5, do 6 as required

3.5.1.2.2.1 position car in centre of lane well clear of fixed barriers

3.5.1.2.2.2 observe general configuration of slip road

3.5.1.2.2.3 observe speed limit if posted

3.5.1.2.2.4 adopt speed appropriate for layout and configuration of off-slip

3.5.1.2.2.5 glance at speedometer to ensure appropriate deceleration/speed

3.5.1.2.2.6 observe route guidance signs

3.5.2 deal with crossroads

Plan 3.5.2 - IF car has right of way AND travelling straight on THEN1 ELSE 2 OR 3

3.5.2.1 traverse crossroad

Plan 3.5.2.1 - WHILE 1 AND 2 do 3 AND 4 AND 5. IF
junction has yellow hatched markings (box junction) THEN 6

   3.5.2.1.1 maintain course

   3.5.2.1.2 maintain speed << GO TO subroutine 2.3.3
'maintain speed' >>

   3.5.2.1.3 observe other traffic

   Plan 3.5.2.1.3 - do 1 AND 2 AND 3 AND 4 AND 5

      3.5.2.1.3.1 observe traffic ahead

      3.5.2.1.3.2 observe oncoming traffic

      3.5.2.1.3.3 observe cross traffic

      3.5.2.1.3.4 observe pedestrians

      3.5.2.1.3.5 watch for any additional hazards

   3.5.2.1.4 generally avoid route changes while in intersection

   3.5.2.1.5 generally avoid stopping in intersection

   3.5.2.1.6 do not stop in intersection

3.5.2.2 turn left

Plan 3.5.2.2 - IF car has to give way AND has stopped at give
way line THEN 1 THEN 2 THEN 3 THEN 4.
IF vehicle has right of way THEN 2 THEN 3
THEN 4

   3.5.2.2.1 check cross traffic

   Plan 3.5.2.2.1 - do 1 THEN 2 THEN 1 THEN 3. IF safe
gap in traffic has arisen THEN 4

      3.5.2.2.1.1 check to the right

      Plan 3.5.2.2.1.1 - WHILE 3 do 1 AND 2

         3.5.2.2.1.1.1 judge distance from nearest vehicle

         3.5.2.2.1.1.2 wait for a gap of sufficient size before
proceeding

         3.5.2.2.1.1.3 generally do not rely on oncoming traffic's
left-hand indications as an intention to turn

      3.5.2.2.1.2 check left-hand blind spot (quickly)

      3.5.2.2.1.3 check that intended course is still clear

      3.5.2.2.1.4 pull away << GO TO subroutine 2.1 'pulling
away from standstill' >>

   3.5.2.2.2 respond to oncoming traffic wanting to turn into
same turning

Plan 3.5.2.2.2 - do 1 ELSE 2 OR 3 to allow safe gap for oncoming driver to drive through THEN 4 OR IF dark OR poor visibility THEN 5 THEN 6. IF vehicle stopped THEN 7

    3.5.2.2.2.1 pass decisively into turn before any oncoming traffic turns

    3.5.2.2.2.2 decelerate << GO TO subroutine 2.4 'decrease speed' >>

    3.5.2.2.2.3 stop

    3.5.2.2.2.4 issue hand signals only to driver of oncoming turning vehicle

    3.5.2.2.2.5 flash vehicles lights twice

    3.5.2.2.2.6 allow oncoming turning vehicle to pass in front of vehicle

    3.5.2.2.2.7 pull away << GO TO subroutine 2.1 'pulling away from standstill' >>

3.5.2.2.3 initiate turn

Plan 3.5.2.2.3 - WHILE 1 AND 2 AND 3 do 4

    3.5.2.2.3.1 avoid swan-necking (turn sharply enough to avoid encroaching upon right lane)

    3.5.2.2.3.2 be careful not to cut corner with vehicles rear wheels

    3.5.2.2.3.3 avoid shifting gears during turn

    3.5.2.2.3.4 feed wheel briskly through hands << GO TO subroutine 2.2.4 'pull-push steering method' >>

3.5.2.2.4 complete turn

Plan 3.5.2.2.4 - WHILE 1 AND 2 do 3 THEN 4

    3.5.2.2.4.1 accelerate slightly during turn

    3.5.2.2.4.2 straighten steering

    3.5.2.2.4.3 cancel indicator

    3.5.2.2.4.4 accelerate to desired speed << GO TO subroutine 2.3.2 'increase current speed' >>

3.5.2.3 turn right

Plan 3.5.2.3 - do 1, IF behind give way line of adjoining road AND IF safe gap in traffic develops THEN 3 THEN 4 THEN 5. IF waiting in middle of road OR middle of crossroad to turn right THEN 1 THEN 5. WHILE 7 AND 8 do 6

3.5.2.3.1 check cross traffic

Plan 3.5.2.3.1 - IF waiting at centre of crossroad THEN 1. IF behind give way line THEN 2 AND 3 AND 4 WHILE 5 AND 6 THEN 7 THEN 8

3.5.2.3.1.1 remain to the right of the centre line

3.5.2.3.1.2 remain behind give way line

3.5.2.3.1.3 keep wheels pointing ahead

3.5.2.3.1.4 keep foot firmly on brake

3.5.2.3.1.5 assess oncoming traffic for a suitable gap

3.5.2.3.1.6 generally do not rely on oncoming vehicle's indications as an intention to turn

3.5.2.3.1.7 check to the rear << GO TO subroutine 5.1.1.2 'rearward surveillance' >>

3.5.2.3.1.8 check to the right to ensure intended path is clear

3.5.2.3.2 check oncoming traffic

Plan 3.5.2.3.2 - do 1 IF oncoming vehicle(s) wish to turn to their right THEN 2 AND (when appropriate) 3 WHILE 4

3.5.2.3.2.1 observe oncoming vehicles wishing to turn to their right

3.5.2.3.2.2 pull alongside oncoming vehicle(s) wishing to turn

3.5.2.3.2.3 pass offside to offside

3.5.2.3.2.4 generally perform manoeuvre decisively

3.5.2.3.3 pull partially into intersection << GO TO 2.1 'pulling away from standstill' >> (briskly)

3.5.2.3.4 begin right turn before reaching centre of crossroad

3.5.2.3.5 turn into left lane of direction of intended travel << GO TO subroutine 2.2.4 'pull-push steering method' >>

3.5.2.3.6 cancel indicator

3.5.2.3.7 straighten steering << GO TO subroutine 2.2.4 'pull-push steering method' >>

3.5.2.3.8 accelerate to desired speed << GO TO subroutine 2.2.4 'pull-push steering method' >>

3.5.3 deal with roundabouts

Plan 3.5.3 - do 1 THEN 2

3.5.3.1 enter roundabout

Plan 3.5.3.1 - WHILE 1 AND 2 AND 3 do 4 IF safe gap in traffic arises THEN 5. IF taking any of the exits that are less than halfway around the roundabout AND unless directed by road signs to the contrary AND/OR unless directed by road markings to the contrary THEN 6. IF taking first exit THEN 9 ELSE 10. IF taking exit that is approximately halfway around roundabout AND unless directed by road signs to the contrary AND/OR unless directed by road markings to the contrary THEN 7 OR 6 WHILE 10. IF taking exits from roundabout that are further than halfway around AND unless directed by road signs to the contrary AND/OR by road markings to the contrary THEN 8 WHILE 10

3.5.3.1.1 give way to traffic approaching from the right that is already on the roundabout

3.5.3.1.2 avoid coming into conflict with other vehicles

Plan 3.5.3.1.2 - do 1 AND 2 AND 3

3.5.3.1.2.1 maintain accurate lane positioning (lane discipline)

3.5.3.1.2.2 maintain safe headway distances

3.5.3.1.2.3 anticipate movements of other vehicles/traffic

3.5.3.1.3 observe directional information

Plan 3.5.3.1.3 - do 1 AND 2

3.5.3.1.3.1 carefully observe road signs providing route information where posted

3.5.3.1.3.2 carefully observe lane markings providing lane and/or route information where posted

3.5.3.1.4 check that path ahead is clear (that any vehicles in front have actually completed the pulling away manoeuvre)

3.5.3.1.5 pull away (briskly) << GO TO subroutine 2.1 'pulling away from standstill' >>

3.5.3.1.6 remain in outside lane of roundabout

3.5.3.1.7 enter middle lane of roundabout

3.5.3.1.8 enter inside lane (unless directed otherwise)

3.5.3.1.9 indicate left

3.5.3.1.10 indicate right

3.5.3.2 leave roundabout

Plan 3.5.3.2 - WHILE 1 AND 2 AND 3 AND IF not taking first left exit AND if not already in outside lane THEN do 4 after passing exit prior to desired exit THEN 5 THEN 6 WHILE 7 THEN 8. IF taking first left exit THEN 5 WHILE 7 THEN 8. IF exit missed OR otherwise unable to have been taken THEN 9 OR 10 and repeat plan (to retrace steps and regain desired route).

3.5.3.2.1 observe movement of other vehicles whilst on roundabout

3.5.3.2.2 << GO TO subroutine 3.5.3.1.2 'avoid coming into conflict with other vehicles' >>

3.5.3.2.3 << GO TO subroutine 3.5.3.1.3 'observe directional information' >>

3.5.3.2.4 indicate left

3.5.3.2.5 check left-hand blind spot (glance left)

3.5.3.2.6 enter outside lane in advance of exit (unless directed otherwise)

3.5.3.2.7 observe traffic entering roundabout during manoeuvre to outside lane

3.5.3.2.8 take exit

3.5.3.2.9 continue around roundabout again

3.5.3.2.10 take next exit

3.5.4 deal with traffic light controlled junctions

Plan 3.5.4 - IF traffic lights/filter arrow red THEN 2. IF traffic lights/filter arrow green AND vehicle in motion THEN 4 ELSE 6. IF traffic lights/filter arrow green for some time THEN 1 WHILE 4. IF traffic lights/filter arrow amber AND IF safe distance to stop THEN 2 ELSE 4. IF traffic lights/filter arrows red and amber simultaneously THEN 5 THEN 6

3.5.4.1 prepare to stop

3.5.4.2 stop vehicle at junction << GO TO subroutine 2.4 'decrease speed' >>

Plan 3.5.4.2 - IF first in queue THEN 1 WHILE 3 AND 4
ELSE 2 WHILE 3 AND 4

3.5.4.2.1 stop behind line

3.5.4.2.2 leave large enough gap to manoeuvre out of behind other queued vehicles

Plan 3.5.4.2.2 - WHILE 1 do 2

3.5.4.2.2.1 avoid pulling up very close to car ahead in queue

3.5.4.2.2.2 ensure that bottom of vehicle ahead's rear tyres are visible, as a guide to a desirable gap

3.5.4.2.3 ensure clear view of traffic lights is available

3.5.4.2.4 observe any vehicles stopping behind

3.5.4.3 wait << GO TO subroutine 2.4.4 'maintain stop (wait)' >>

3.5.4.4 proceed through junction (decisively)

Plan 3.5.4.3 - do 1 AND 2

3.5.4.4.1 observation << GO TO subroutine 5.1 'surveillance' >>

3.5.4.4.2 be prepared to stop

3.5.4.5 prepare to pull away

3.5.4.6 pull away << GO TO subroutine 2.1 'pulling away from standstill' >>

3.6 deal with crossings

Plan 3.6 - WHILE 5 do 1 OR 2 OR 3 OR 4

3.6.1 deal with zebra crossings (Belisha beacons)

Plan 3.6.1 - do 1 IF people exhibiting desire to use crossing THEN 2 THEN 3 WHILE 4. IF pedestrians already on crossing THEN 3 WHILE 4

3.6.1.1 observe for people at the side of the road wanting to use crossing

3.6.1.2 << GO TO subroutine 2.4 'decrease speed' >>

3.6.1.3 give way to pedestrians

3.6.1.4 do not wave pedestrians across crossing

3.6.2 deal with pelican crossings (traffic lights)

Plan 3.6.2 - do 1 (the only difference between normal traffic lights is that pelican crossings flash the amber light after the red phase) IF amber light flashing AND the crossing is not being used/finished being used AND the way ahead is clear THEN 2. IF traffic island included in crossing THEN 3

3.6.2.1 << GO TO subroutine 3.5.4 'deal with traffic light controlled junctions' >>

3.6.2.2 proceed with caution

3.6.2.3 treat as one crossing

3.6.3 deal with toucan crossings

Plan 3.6.3 - do 1 AND 2 (crossing as per any other traffic light controlled junction with no flashing amber phase, permits cyclists to ride across as well as pedestrians)

3.6.3.1 << GO TO subroutine 3.5.4 'deal with traffic light controlled junctions' >>

3.6.3.2 check for cyclists emerging from side of junction

3.6.4 deal with railway crossings

Plan 3.6.4 - IF flashing red lights AND audible warning AND barriers being lowered THEN 1. IF amber light flashing AND audible warning sounding AND unsafe to stop THEN 2 WHILE 5. IF train has passed AND barriers remain lowered AND lights remain flashing AND audible warning still sounds THEN 3. IF audible warning stops AND barriers rise AND lights stop flashing THEN 4 WHILE 5

3.6.4.1 << GO TO subroutine 2.4 'decrease speed' >>

3.6.4.2 proceed decisively across tracks

3.6.4.3 continue waiting

3.6.4.4 << GO TO subroutine 2.1 'pulling away from stand-still' >>

3.6.4.5 do not stop on tracks (ensure way ahead is clear)

3.6.5 perform general crossing tasks

Plan 3.6.5 - WHILE 1 AND 2 AND 3 do 4

3.6.5.1 never stop on crossing

3.6.5.2 observe movements of people/animals/vehicles that look like they will be/or desire to use the crossing

3.6.5.3 << GO TO subroutine 5.1 'surveillance' >>

3.6.5.4 obey lights

Plan 3.6.5.1 - IF lights red THEN 1. IF lights amber THEN 1 ELSE 2. IF lights green THEN 3

3.6.5.1.1 stop

3.6.5.1.2 proceed across crossing if unsafe to stop

3.6.5.1.3 proceed across crossing with due caution

3.7 leave junction (crossing)

Plan 3.7 - WHILE 4 AND 5 do 1 AND 2 AND 3

   3.7.1 cancel indicators << GO TO subroutine 2.1 'pulling away from standstill' >>

   3.7.2 observe road signs for further route/lane guidance

   3.7.3 observe road markings for further route/lane guidance

   3.7.4 adjust speed

   Plan 3.7.4 - WHILE 1 OR 2 AND 3 do 4 OR 5 OR 6

      3.7.4.1 observe posted speed limit

      3.7.4.2 deduce from road type the appropriate speed limit << GO TO subroutine 5.3 'rule compliance' >>

      3.7.4.3 gauge relative speed of traffic flow

      3.7.4.4 increase speed << GO TO subroutine 2.3.2 'increase current speed' >>

      3.7.4.5 decrease speed << GO TO subroutine 2.4 'decrease current speed' >>

      3.7.4.6 maintain existing speed << GO TO subroutine 2.3.3 'maintain speed' >>

   3.7.5 resume normal/pre-junction driving

4: PERFORM TACTICAL DRIVING TASKS

Plan 4 - do 1 AND 2 AND 3. do 4 as required

4.1 deal with different road types/classifications

Plan 4.1 - IF driving on urban roads OR encountering urban road features in other settings THEN 1. IF driving on rural roads OR encountering rural road features in other settings THEN 2. IF driving on main roads OR encountering main road features in other settings THEN 3. IF driving on motorways OR encountering motorway features in other settings THEN 4

   4.1.1 drive in urban settings

   Plan 4.1.1 - WHILE 6, do 1 AND 2 AND 3 AND 4 AND 5

      4.1.1.1 observe other road users in urban setting

      Plan 4.1.1.1 - WHILE 1 IF movements of other road users necessitate THEN 2 as required

         4.1.1.1.1 observe movements of other road users

Plan 4.1.1.1.1 - WHILE 4 do 1 AND/OR 2 AND/OR 3
THEN 5 based on observation of relevant features

4.1.1.1.1.1 observe other vehicles

Plan 4.1.1.1.1.1 - do 1 AND 2 AND 3 AND 4 as relevant features arise

4.1.1.1.1.1.1 observe rows of parked vehicles

4.1.1.1.1.1.2 observe buses waiting at bus stops

4.1.1.1.1.1.3 observe trade vehicles

4.1.1.1.1.1.4 observe for ice cream vans/mobile shops/
school buses etc.

4.1.1.1.1.2 observe movements of cyclists

Plan 4.1.1.1.1.2 - do 1 AND 2 AND 3 AND 4 as relevant features arise

4.1.1.1.1.2.1 observe inexperienced cyclists doing anything erratic

4.1.1.1.1.2.2 observe roadroad sside hazards that may
cause cyclist(s) to swerve

4.1.1.1.1.2.3 cyclist(s) looking over right shoulder

4.1.1.1.1.2.4 observe any strong wind/gusts

4.1.1.1.1.2.5 observe young cyclists doing anything
dangerous (wheelies etc.)

4.1.1.1.1.3 observe pedestrians

Plan 4.1.1.1.1.3 - do 1 AND 2 as relevant circumstance
arises

4.1.1.1.1.3.1 observe movement of pedestrians

4.1.1.1.1.3.2 observe pedestrian actions (such as hailing
taxi/looking both ways)

4.1.1.1.1.4 observe general urban road environment

Plan 4.1.1.1.1.4 - do 1 AND 2 AND 3AND 4 as relevant
road/environment present themselves

4.1.1.1.1.4.1 observe side turnings

4.1.1.1.1.4.2 observe pull-ins/petrol station entrances/
parking bays etc.

4.1.1.1.1.4.3 observe pull-outs/petrol station exits/
parking bays etc.

4.1.1.1.1.4.4 generally anticipate/prepare for any road
users coming into conflict with desired speed/trajectory of vehicle

4.1.1.1.2 deal with movements of other road users

Plan 4.1.1.1.2 - IF changes in speed will avert coming into conflict/collision with other road users THEN 1 OR 2 ELSE 3

    4.1.1.1.2.1 increase speed << GO TO subroutine 2.3.2 'increase speed' >>

    4.1.1.1.2.2 decrease speed << GO TO subroutine 2.4 'decrease speed' >>

    4.1.1.1.2.3 take evasive action << GO TO subroutine 4.4.1 'take evasive action' >>

4.1.1.2 observe road signs

Plan 4.1.1.2 - WHILE 1 AND 2 do 3

    4.1.1.2.1 observe signs indicating speed limit for road

    4.1.1.2.2 observe signs denoting hazards ahead

    4.1.1.2.3 observe signs providing route information

4.1.1.3 observe lane markings

4.1.1.4 anticipate junctions/crossings/intersections

4.1.1.5 deal with traffic calming measures

4.1.1.6 lane usage

Plan 4.1.1.6 - WHILE 1 AND 2 AND 3 IF needing to pass vehicle(s) ahead THEN 4 ELSE 5 AND use this lane to pass

    4.1.1.6.1 drive in far left lane

    4.1.1.6.2 position car in centre of lane

    4.1.1.6.3 attempt to stay in lane as far as possible

    4.1.1.6.4 use right lane(s) to pass

    4.1.1.6.5 assume position on road corresponding to the appropriate travel lane

4.1.2 drive in rural settings

Plan 4.1.2 - WHILE 1 AND 2 AND 3 AND 4 do 5 THEN 6 as required

    4.1.2.1 observation in rural driving

Plan 4.1.2.1 - do 1 AND 2 AND 3 AND 4 AND 5 as relevant features arise

    4.1.2.1.1 observe other vehicles in rural setting

Plan 4.1.2.1.1 - do 1 AND/OR 2 AND/OR 3 as required

        4.1.2.1.1.1 observe for large vehicles on narrow roads

        4.1.2.1.1.2 observe for slow moving agricultural machinery

4.1.2.1.1.3 observe for slow moving/indecisive tourist traffic

4.1.2.1.2 observe other road users in rural setting

Plan 4.1.2.1.2 - do 1 AND 2 as situation arises

    4.1.2.1.2.1 observe for hikers/walkers

    4.1.2.1.2.2 observe for leisure road users (such as horse riders/cyclists etc.)

4.1.2.1.3 observe pertinent road features

Plan 4.1.2.1.3 - do 1 as indication that junction is ahead, do 2 as indication of side turning ahead, do 3 as indication that horses/farm machinery etc. may be encountered ahead soon AND that road may be slippery

    4.1.2.1.3.1 observe for clusters of lamp posts ahead

    4.1.2.1.3.2 observe for single lamp posts on road side

    4.1.2.1.3.3 observe for fresh mud or other deposits on road

4.1.2.1.4 observe road signs

Plan 4.1.2.1.4 - do 1 AND/OR 2

    4.1.2.1.4.1 observe road signs indicating curvature of road

    4.1.2.1.4.2 observe road signs indicating gradient of road

4.1.2.1.5 observe for animals in the road

4.1.2.2 anticipate course of road

Plan 4.1.2.2 - do 1 AND 2 AND 3 as indications that road is going to change direction (and possibly in what direction)

    4.1.2.2.1 observe whether there is a gap in trees ahead of road

    4.1.2.2.2 check whether road is running alongside railway line

    4.1.2.2.3 observe for natural barriers (such as rivers/large hills etc.)

4.1.2.3 use road lanes appropriately

Plan 4.1.2.3 - unless directed otherwise WHILE 1 do 2 IF approaching vehicles AND road is narrow THEN 3. IF still not sufficient passing room THEN 4 AND 5 if necessary

    4.1.2.3.1 maintain car in centre of driving lane

    4.1.2.3.2 drive in left lane

4.1.2.3.3 move over to left of lane

4.1.2.3.4 pull off roadway

4.1.2.3.5 stop << GO TO subroutine 2.4 'decrease speed' >>

4.1.2.4 observe road markings

4.1.2.5 anticipate/prepare for any road users coming into conflict with desired speed/trajectory of vehicle

4.1.2.6 << GO TO subroutine 4.1.1.1.2 'deal with movements of other road users' >>

4.1.3 drive on a main road

Plan 4.1.3 - WHILE 1 do 2 AND 3

4.1.3.1 << GO TO subroutine 4.1.2 'rural driving' >>

4.1.3.2 deal with other vehicles in main road setting

Plan 4.1.3.2 - WHILE 1 do 2 AND 3 AND 4. IF hazard presents itself AND needs to be passed THEN 5 AND 6 ELSE 7 as required

4.1.3.2.1 anticipate large differentials in speed (due to the wide range of vehicles permitted to use main roads)

4.1.3.2.2 observe for overtaking vehicles

4.1.3.2.3 observe for vehicles emerging from junctions/side turnings

4.1.3.2.4 observe for cyclists

4.1.3.2.5 move over to the centre of the road

4.1.3.2.6 pass on the right any slower moving road user

4.1.3.2.7 << GO TO subroutine 3.3 'overtaking' >>

4.1.3.3 deal with main road features

Plan 4.1.3.3 - do 1 AND 2 AND 3 AND 4

4.1.3.3.1 respond to changes in speed limits

4.1.3.3.2 respond to changes in main road environment (open road/urban settings)

Plan 4.1.3.3.2 - do 1 AND/OR 2 AND/OR 3 AND/OR 4

4.1.3.3.2.1 plan need to adjust following distances

4.1.3.3.2.2 plan opportunities for overtaking

4.1.3.3.2.3 plan need to adjust road position/speed for increased visibility

4.1.3.3.2.4 respond appropriately to general road situation/hazards

4.1.3.3.3 respond to changes in road markings

4.1.3.3.4 respond to changes in lane structure

4.1.4 drive on a motorway or dual carriageway

Plan 4.1.4 - WHILE 1 AND 2 IF slower moving traffic OR any
 other road user/hazard requires it THEN 3

  4.1.4.1 consider road/traffic conditions

  Plan 4.1.4.1 - do 1 AND 2 AND 3 AND 4 AND 5

    4.1.4.1.1 anticipate likely traffic volume

    4.1.4.1.2 anticipate the possibility of roadworks or other delays

    4.1.4.1.3 assess road/traffic conditions well ahead

    4.1.4.1.4 plan manoeuvres well in advance

    4.1.4.1.5 observe for specific motorway (dual carriageway)
 hazards

    Plan 4.1.4.1.5 - do 1 AND 2 AND 3

      4.1.4.1.5.1 observe for other vehicle's lane change
 manoeuvres coming into conflict with cars desired
 speed and trajectory

      4.1.4.1.5.2 observe for vehicles leaving motorway exit
 manoeuvres too late

      4.1.4.1.5.3 generally attempt to avoid overtaking three
 abreast, or otherwise leaving no room for evasive
 action/manoeuvre

  4.1.4.2 exhibit lane discipline

  Plan 4.1.4.2 - do 1 AND 2. IF traffic queuing THEN 3 IF
 car's traffic queue is moving faster

    4.1.4.2.1 generally drive in left-hand lane

    4.1.4.2.2 generally only overtake on the right

    4.1.4.2.3 pass queuing traffic using left lanes

  4.1.4.3 passing/overtaking on motorways (dual carriageways)

  Plan 4.1.4.3 - do 1 IF overtaking opportunity arises THEN 2
 THEN 3 THEN 4 THEN 5

    4.1.4.3.1 undertake information phase

    Plan 4.1.4.3.1 - do 1 AND 2 AND 3 AND 4 AND 5

      4.1.4.3.1.1 observe for slower moving vehicles moving out
 in front of vehicle

      4.1.4.3.1.2 observe for faster moving vehicles approaching
 from behind

      4.1.4.3.1.3 observe relative speeds of other drivers

      4.1.4.3.1.4 observe head/body movements of other drivers

4.1.4.3.1.5 observe for vehicle movement from the centre of the lane towards the white lane markers

4.1.4.3.2 undertake speed phase

Plan 4.1.4.3.2 - IF overtaking opportunity requires THEN 1 OR 2 ELSE 3

4.1.4.3.2.1 << GO TO subroutine 2.3.2 'increase speed' >>

4.1.4.3.2.2 << GO TO subroutine 2.3.3 'maintain speed' >>

4.1.4.3.2.3 << GO TO subroutine 2.4 'decrease speed' >>

4.1.4.3.3 give information to other road users

Plan 4.1.4.3.3 - do 1 IF warning of approach required for drivers ahead THEN 2

4.1.4.3.3.1 activate left/right indicator long enough for drivers to react to it

4.1.4.3.3.2 provide extended flash of headlights

4.1.4.3.4 move into new lane

Plan 4.1.4.3.4 - WHILE 1 do 2 AND 3

4.1.4.3.4.1 ensure lane change is completed decisively (progressively)

4.1.4.3.4.2 << GO TO subroutine 3.3.5.1.5 'recheck what is happening behind' >>

4.1.4.3.4.3 avoid harsh/abrupt manoeuvres

Plan 4.1.4.3.4.3 - do 1 THEN 2 and allow vehicle to begin crossing lane dividing lines. IF both front wheels have crossed the central lane divide THEN 3 until normal/central lane position is achieved THEN 2

4.1.4.3.4.3.1 perform initial steering input (very gently)

4.1.4.3.4.3.2 return steering to straight ahead

4.1.4.3.4.3.3 perform corrective steering input (very gently)

4.1.4.3.5 cancel indicators

4.2 deal with roadway-related hazards

Plan 4.2 - do 1 AND/OR 2 AND/OR 3

4.2.1 deal with different types of road surface

Plan 4.2.1 - do 1 AND 2 AND 3

4.2.1.1 observe nature of road surface materials upon which car is being driven

4.2.1.2 adjust movements of car to nature of road surface

Plan 4.2.1.2 - do 1 AND 2 AND 3 AND 4

    4.2.1.2.1 drive more slowly than on dry paved/metaled road

    4.2.1.2.2 avoid sharp turning movements

    4.2.1.2.3 generally avoid sharp braking actions

    4.2.1.2.4 increase following distances

4.2.1.3 observe for conditions specific to type of road surface materials

Plan 4.2.1.3 - IF on normal roads THEN 1. IF driving on unmade roads THEN 2. IF driving on gravel THEN 3. IF driving on cobbles/bricks THEN 4

    4.2.1.3.1 anticipate smoothness of concrete or asphalt road surface

    4.2.1.3.2 check for loose soil conditions and hazardous objects such as rocks, glass, sharp objects embedded in road

    4.2.1.3.3 check for loose gravel

    4.2.1.3.4 check for holes, bumps, cracks, loose bricks and slippery spots

4.2.2 deal with road surface irregularities

Plan 4.2.2 - do 1 IF road condition deficient THEN 2 IF particularly harsh road conditions detected OR potholes THEN 3 as necessary

    4.2.2.1 observe road surface for surface defects and irregularities caused by weather and/or general road deterioration

    4.2.2.2 reduce car speed

    4.2.2.3 avoid/mitigate effects of wheels hitting pot holes

Plan 4.2.2.3 - WHILE 4, do 1 ELSE 2 THEN 3

    4.2.2.3.1 reposition car to straddle pothole

    4.2.2.3.2 reduce speed << GO TO subroutine 2.4 'decrease speed' >>

    4.2.2.3.3 release brake as wheel descends into pothole (so that suspension is fully unloaded, and more of the suspension's travel is available to soak up bump/rebound).

    4.2.2.3.4 grip wheel firmly

4.2.3 deal with obstructions

Plan 4.2.3 - do 1 AND/OR 2

4.2.3.1 deal with objects in road

Plan 4.2.3.1 - do 1 AND 2 as required

 4.2.3.1.1 observe for hazardous objects

 Plan 4.2.3.1.1 - do 1 AND 2 AND 3

  4.2.3.1.1.1 check for puddles, rivulets, particularly where drainage is poor

  4.2.3.1.1.2 check for rock slides and debris

  4.2.3.1.1.3 check for other debris

 4.2.3.1.2 respond to hazardous objects

 Plan 4.2.3.1.2 - WHILE 1 AND 2 do 3 as required

  4.2.3.1.2.1 maintain slower speed until road area is clear of hazardous objects

  4.2.3.1.2.2 do not come into conflict with other/oncoming traffic

  4.2.3.1.2.3 << GO TO subroutine 4.4.1 'take evasive action' >>

4.2.3.2 deal with roadworks and barricades

Plan 4.2.3.2 - do 1 AND 2 AND 3 AND 4

 4.2.3.2.1 observe for indications/signs denoting roadworks

 4.2.3.2.2 drive at reduced speed

 4.2.3.2.3 prepare to stop if necessary

 4.2.3.2.4 maintain increased alertness to the movements of people and machinery

4.3 react to other traffic

Plan 4.3 - do 1 AND 2 AND 3 if required

 4.3.1 reacting to other vehicles

 Plan 4.3.1 - do 1 AND/OR 2 THEN 3 AND/OR 4 AND/OR 5 AND/OR 6 as required

  4.3.1.1 reacting to parked vehicles

  Plan 4.3.1.1 - IF approaching OR driving alongside parked vehicles THEN 1 AND 2 AND 3 AND 4 THEN 5 IF lead vehicle(s) about to enter OR exit a parking space OR in response to animals/pedestrians/vehicle doors being opened/people emerging between parked vehicles THEN 6 OR 7 as required

   4.3.1.1.1 drive at slower speeds when approaching or driving alongside parked vehicles

4.3.1.1.2 observe for pedestrians or animals entering the road from in front of, or between parked cars

4.3.1.1.3 observe for vehicle doors being opened

4.3.1.1.4 observe for vehicles about to pull out from roadside

Plan 4.3.1.1.4 - do 1 AND 2 AND 3 AND 4

4.3.1.1.4.1 observe for vehicles with drivers sitting inside

4.3.1.1.4.2 observe vehicles with engine running as evidenced by exhaust smoke

4.3.1.1.4.3 observe for indicators, tail lights, or stop lights

4.3.1.1.4.4 observe for vehicles where front wheels are being steered outwards

4.3.1.1.5 provide indication/warning of car's presence on road

Plan 4.3.1.1.5 - do 1 AND/OR 2

4.3.1.1.5.1 sound horn

4.3.1.1.5.2 flash headlights

4.3.1.1.6 prepare to stop behind/change lane when vehicle ahead is about to exit or enter a parking space

Plan 4.3.1.1.6 - IF vehicle in process of parking WHILE 1 AND 2 THEN 3 as required AND 4. IF other vehicle is parallel parking THEN 5

4.3.1.1.6.1 allow sufficient clearance ahead to enable the vehicle driver to complete their manoeuvre without crowding

4.3.1.1.6.2 make certain driver of parked/parking vehicle is aware of vehicles presence << GO TO 4.3.1.1.5 'provide indication/warning of vehicle's presence on road' >>

4.3.1.1.6.3 change lane with appropriate caution

4.3.1.1.6.4 make sure there is adequate clearance ahead

4.3.1.1.6.5 allow a full car width between car and vehicle that is parallel parking

4.3.1.1.7 << GO TO subroutine 4.4.1 'take evasive action' >>

4.3.1.2 reacting to being followed

Plan 4.3.1.2 - IF decelerating AND/OR stopping THEN WHILE 1 do 2 AND 3. IF wishing to change direction THEN 1 WHILE 3 (periodically). IF following vehicles passing OR overtaking THEN 4 ELSE 5 WHILE 6

4.3.1.2.1 clearly signal intentions to following driver

Plan 4.3.1.2.1 - do 1 AND 2

    4.3.1.2.1.1 use indicators well in advance of manoeuvres

    4.3.1.2.1.2 deploy brake lights

4.3.1.2.2 make smooth and gradual stops

Plan 4.3.1.2.2 - do 1 AND 2

    4.3.1.2.2.1 observe road/traffic ahead to anticipate stop requirements

    4.3.1.2.2.2 decelerate early and (progressively)

4.3.1.2.3 check rear-view mirror (frequently)

4.3.1.2.4 observe rate of overtaking by following vehicle

4.3.1.2.5 observe following vehicle's indicators for intent to pass

4.3.1.2.6 watch for tailgating vehicles

Plan 4.3.1.2.6 - WHILE 4 do 1 ELSE 2 to encourage vehicle to pass. IF travelling in passing lane of dual carriageway/motorway THEN 3

    4.3.1.2.6.1 maintain speed

    4.3.1.2.6.2 gradually slow down

    4.3.1.2.6.3 return to nearside lane(s) at safe opportunity

    4.3.1.2.6.4 avoid abrupt reactions

4.3.1.3 responding to being passed

Plan 4.3.1.3 - do 1 AND 2 AND 3 AND 4 ELSE 5 AND 6 AND 7 AND 8 IF passing vehicle experiencing problems THEN 9

    4.3.1.3.1 check rear-view mirror frequently

    4.3.1.3.2 use peripheral vision to detect overtaking/passing vehicles

    4.3.1.3.3 check ahead to determine whether other vehicle's pass can be safely completed

    4.3.1.3.4 maintain centre lane position

    4.3.1.3.5 adjust position slightly to provide additional passing clearance

    4.3.1.3.6 maintain speed (do not accelerate)

    4.3.1.3.7 watch for signals or other indications that the passing vehicle plans to cut back in front

    4.3.1.3.8 prepare to decelerate to provide larger opening for passing car to slot into after passing

4.3.1.3.9 respond to problems related to passing vehicle

Plan 4.3.1.3.9 - WHILE 2 AND 3, IF passing vehicle having difficulty completing manoeuvre THEN 1 OR 5 as necessary to allow passing vehicle to slot into left lane with minimum difficulty ELSE 4

    4.3.1.3.9.1 decelerate as necessary

    4.3.1.3.9.2 plan 'escape route'

    4.3.1.3.9.3 maintain grip on steering wheel

    4.3.1.3.9.4 << GO TO subroutine 4.4.1 'take evasive action' >> using 'escape route'

    4.3.1.3.9.5 accelerate quickly to allow passing driver to pull back in behind

4.3.1.4 react to oncoming vehicles

Plan 4.3.1.4 - WHILE 1 AND 2 AND 3 AND 4 AND 5. IF collision with oncoming vehicle imminent THEN 6

    4.3.1.4.1 generally use left lane where possible

    4.3.1.4.2 maintain precise control over car when passing oncoming vehicles

    4.3.1.4.3 react quickly to wind gusts, road irregularities, etc.

    4.3.1.4.4 observe for indication that oncoming vehicle might cross centre line

    Plan 4.3.1.4.4 - do 1 AND 2 AND 3 AND 4 AND 5

        4.3.1.4.4.1 observe turn signals of approaching vehicles

        4.3.1.4.4.2 observe oncoming tailgating vehicles, suggesting desire to pass

        4.3.1.4.4.3 observe slow moving or stopped vehicles in anticipation of vehicles pulling out to pass

        4.3.1.4.4.4 observe vehicles pulling/backing out of parking spaces

        4.3.1.4.4.5 watch for drivers cutting/drifting across centre line on curves

        4.3.1.4.4.6 observe for 'lurkers' stuck behind large vehicles ahead

    4.3.1.4.5 observe roadway for conditions that might cause oncoming vehicle to stray across into vehicles lane

Plan 4.3.1.4.5 - do 1 AND 2 AND 3 AND 4

    4.3.1.4.5.1 observe road for slippery surface

    4.3.1.4.5.2 watch for ruts

    4.3.1.4.5.3 watch for potholes

    4.3.1.4.5.4 watch for other obstructions (debris, road-works etc.)

4.3.1.4.6 react to oncoming traffic on collision course

Plan 4.3.1.4.6 - do 1 THEN 2 THEN 3 if necessary

    4.3.1.4.6.1 reduce speed or stop << GO TO subroutine 2.4 'decrease speed' >>

    4.3.1.4.6.2 signal to other driver << GO TO subroutine 4.3.1.1.5 'provide indication/warning of car's presence on road' >>

    4.3.1.4.6.3 take evasive/avoiding action << GO TO subroutine 4.4.1 'take evasive action' >>

4.3.1.5 react to vehicle ahead

Plan 4.3.1.5 - do 1, IF car speed > lead vehicle speed THEN 2 THEN 3 ELSE IF road/traffic situations permit AND need OR desire exists THEN 4 OR 5 ELSE 6

4.3.1.5.1 determine closing rate of car with lead vehicle

Plan 4.3.1.5.1 - WHILE 1, do 2 AND 3 AND 4

    4.3.1.5.1.1 judge closing rate

    4.3.1.5.1.2 anticipate typically slow moving vehicles such as farm machines, trucks on hills etc.

    4.3.1.5.1.3 anticipate frequently stopping vehicles such as buses, post vans etc.

    4.3.1.5.1.4 anticipate vehicles that are engaged in turning, exiting/entering road, approaching crossings etc.

    4.3.1.5.2 << GO TO subroutine 2.4 'decelerate' >>

    4.3.1.5.3 << GO TO subroutine 3.2 'following other vehicles' >>

    4.3.1.5.4 pass vehicle

    4.3.1.5.5 << GO TO subroutine 3.3 'overtaking' >>

    4.3.1.5.6 reduce speed and operate independently of lead vehicle

4.3.1.6 react to special vehicles

Plan 4.3.1.6 - IF buses encountered THEN 1. IF police/ emergency vehicles encountered THEN 2 ELSE WHILE 3 do 4 AND 5 OR 6 OR 7 OR 8 as required

4.3.1.6.1 react to buses

Plan 4.3.1.6.1 - do 1. IF bus is stopping THEN 2,3. WHILE 4 do 5

4.3.1.6.1.1 look for indication that bus is about to stop

Plan 4.3.1.6.1.1 do 1 AND 2 AND 3 AND 4 AND 5

4.3.1.6.1.1.1 look for indicators

4.3.1.6.1.1.2 look for brake lights

4.3.1.6.1.1.3 look for bus passengers beginning to stand and make their way to the front

4.3.1.6.1.1.4 look for groups of people at bus stops

4.3.1.6.1.1.5 look for bus stop signs

4.3.1.6.1.2 come to a complete halt at safe distance behind bus << GO TO subroutine 2.4 'decrease speed' >>

4.3.1.6.1.3 remain stopped << GO TO subroutine 2.4.4 'maintain stop - wait' >>

4.3.1.6.1.4 observe for passengers who have just alighted at the side of the road

4.3.1.6.1.5 set off << GO TO subroutine 2.1 'pulling away from standstill' >>

4.3.1.6.2 react to police/emergency vehicles

Plan 4.3.1.6.2 - IF emergency vehicle not seen but heard THEN 1 AND 2. IF emergency vehicle seen/localized WHILE 3 THEN 4 THEN 5 OR 6. IF behind emergency vehicle THEN 8 WHILE 7 ELSE 9

4.3.1.6.2.1 localize siren sounds

4.3.1.6.2.2 exercise extreme caution when crossing intersections/junctions

4.3.1.6.2.3 do not cause an obstruction

4.3.1.6.2.4 pull over to side and stop

4.3.1.6.2.5 proceed only when sure that emergency vehicle has passed

4.3.1.6.2.6 remain stopped and await further instructions from police officer/fireman/ambulance crew etc.

4.3.1.6.2.7 exercise extreme caution

4.3.1.6.2.8 prepare to stop

4.3.1.6.2.9 follow emergency vehicle at distance <500feet

4.3.1.6.3 maintain safe distance

4.3.1.6.4 plan need to adjust speed/headway/trajectory of vehicle

4.3.1.6.5 << GO TO subroutine 2.4 'decrease speed' >>

4.3.1.6.6 << GO TO subroutine 3.2 'follow other vehicles' >>

4.3.1.6.7 pass other vehicle(s)

4.3.1.6.8 << GO TO subroutine 3.3 'overtaking' >>

4.3.2 respond to pedestrians and other road users

Plan 4.3.2 - 1 AND 2 AND 3

4.3.2.1 observe pedestrians

Plan 4.3.2.1 - do 1 AND 2 AND 3 AND 4

4.3.2.1.1 observe for pedestrians near intersections, pelican and zebra crossings

4.3.2.1.2 check for indication that pedestrian is to cross in path of car

4.3.2.1.3 observe for jay walkers/people who are running or distracted

4.3.2.1.4 observe for children

4.3.2.2 pass pedestrians

Plan 4.3.2.2 - WHILE 1 do 2. do 3 as necessary to warn pedestrians of vehicle's approach

4.3.2.2.1 prepare to stop

4.3.2.2.2 provide maximum clearance when passing pedestrians

4.3.2.3.3 sound horn

4.3.2.3 watch out for animals (domestic and wildlife) in road

Plan 4.3.2.3 - WHILE 3 do 1 AND IF animal unaware of vehicles approach THEN 2 as required. IF safe to pass animal THEN 4

4.3.2.3.1 decelerate when entering animal crossing zones or when noting animals on/alongside roadway

4.3.2.3.2 sound horn to alert animals of vehicle's approach

4.3.2.3.3 prepare to stop or swerve if animal enters road << GO TO subroutine 4.4.1 'take evasive action' >>

4.3.2.3.4 overtake/pass animal

Plan 4.3.2.3.4 - do 1 AND 2

   4.3.2.3.4.1 provide large passing clearance

   4.3.2.3.4.2 avoid creating excessive noise

   Plan 4.3.2.3.4.2 - do 1 AND 2 (except if animal is particularly obstinate)

      4.3.2.3.4.2.1 avoid racing engine

      4.3.2.3.4.2.2 generally avoid sounding horn

4.3.2.4 observe cyclists

Plan 4.3.2.4 - WHILE 1 do 2 AND 3 THEN 4 as necessary to alert cyclists

   4.3.2.4.1 judge speed carefully (cycles can reach 30+MPH)

   4.3.2.4.2 watch for young cyclists

   4.3.2.4.3 watch for unconfident/unsteady/careless/reckless cyclists

   4.3.2.4.4 sound horn

4.3.2.5 pass cyclists

Plan 4.3.2.5 - do 1 THEN 2 THEN 3 OR 4 as appropriate

   4.3.2.5.1 << GO TO subroutine 5.1 'surveillance' >>

   4.3.2.5.2 indicate

   4.3.2.5.3 pass cyclist leaving plenty of room

   4.3.2.5.4 << GO TO subroutine 3.3 'overtaking' >>

4.3.2.6 respond appropriately to motorcyclists

Plan 4.3.2.6 - do 1 AND 2 IF motorcycle showing indications of passing THEN 3

   4.3.2.6.1 check relative speed very carefully

   4.3.2.6.2 pay particular care with observation

   4.3.2.6.3 adjust position in lane to allow motorbikes to pass safely

4.3.3 react to accident/emergency scenes

Plan 4.3.3 - do in order

   4.3.3.1 approach scene of accident or emergency

   Plan 4.3.3.1 - WHILE 1 do 2 AND 3 AND 4

      4.3.3.1.1 prepare to stop if required

      4.3.3.1.2 slow down in advance of affected area

      4.3.3.1.3 observe for traffic officers or other persons at the scene

    4.3.3.1.4 observe for indications or instructions regarding car movement through affected area

Plan 4.3.3.1.4 - do 1 AND 2. IF approaching scene of accident in immediate aftermath AND scene is unattended THEN 3 THEN 4

    4.3.3.1.4.1 check for signals by persons stationed at the scene controlling traffic movement

    4.3.3.1.4.2 look for signs/cones or other warning devices outlining the route through the area

    4.3.3.1.4.3 stop at scene of accident in safe location (completely off road if possible)

    4.3.3.1.4.4 provide assistance as required

  4.3.3.2 drive by or through emergency area

Plan 4.3.3.2 - do 1 AND 2 AND 3 THEN 4

    4.3.3.2.1 drive at reduced speed

    4.3.3.2.2 watch for unexpected movement of vehicles and pedestrians on the road

    4.3.3.2.3 do not 'rubberneck' (slow down or stop unnecessarily to view emergency scene activities)

    4.3.3.2.4 resume normal speed only after completely passing the emergency area

4.4 perform emergency manoeuvres

Plan 4.4 - do 1 OR 2

  4.4.1 take evasive action

Plan 4.4.1 - WHILE 1 AND 2 do 3 AND/OR 4 AND/OR 5 as required to avoid collision. IF collision cannot be avoided THEN 6

    4.4.1.1 grip wheel firmly

    4.4.1.2 consider steering/cornering/braking grip trade-off

    4.4.1.3 operate wheel (vigorously) as required to complete manoeuvre

    4.4.1.4 swerve an amount sufficient to avoid collision

Plan 4.4.1.4 - WHILE 1 AND 2 IF manoeuvre occurring at speeds >40mph THEN 3 OR 4 as required to complete manoeuvre ELSE IF speed <40mph THEN generally avoid performing 3

    4.4.1.4.1 steer into planned 'escape route'

4.4.1.4.2 bring vehicle to limits of lateral grip as required to complete manoeuvre

Plan 4.4.1.4.2 - do 1 AND 2 IF tyres chirping AND aligning torque decreasing THEN cornering limits have been approached. IF tyres screech continuously (loudly) OR steering aligning torque has disappeared THEN 3

4.4.1.4.2.1 check for sound of 'chirping' tyres

4.4.1.4.2.2 check (via haptics) rapid decline in steering aligning torque

4.4.1.4.2.3 << GO TO subroutine 4.4.2.3 'skid detection' >>

4.4.1.4.3 maintain hands at quarter to 3 position on steering wheel rim

4.4.1.4.4 << GO TO subroutine 2.2.5 'rotational steering' >>

4.4.1.5 brake an amount sufficient to avoid collision

Plan 4.4.1.5 - WHILE 1 do 2

4.4.1.5.1 build up pressure on brake pedal (quickly and progressively)

4.4.1.5.2 attempt to maintain vehicle's wheels at brink of locking

Plan 4.4.1.5.2 - do 1 IF 'chirping' heard THEN 3 until stopped IF speed >40mph AND/OR travelling in a relatively straight line THEN 2 WHILE 3 ELSE 4

4.4.1.5.2.1 listen for intermittent 'chirping' of tyres on road (as opposed to continuous screeching/scrubbing)

4.4.1.5.2.2 ignore locked rear wheels

4.4.1.5.2.3 maintain pressure on brake pedal

4.4.1.5.2.4 << GO TO subroutine 4.4.2.4.2 'cadence braking' >>

4.4.1.6 collision mitigation

Plan 4.4.1.6 - WHILE 1 do 2

4.4.1.6.1 attempt to hit obstacles head-on (vehicles offer best crash protection in longitudinal plane)

4.4.1.6.2 brace for collision

4.4.2 control skids

Plan 4.4.2 - do steps 1 to 3 in order IF skid/slide detected THEN 4 ELSE 5

4.4.2.1   anticipate skid producing situations

Plan 5.4.2.1 - do 1 AND 2 AND 3 AND 4 AND 5 AND 6

    4.4.2.1.1   avoid speed which is excessive for the road conditions

    4.4.2.1.2   avoid acceleration which is excessive for the road conditions

    4.4.2.1.3   avoid excessive braking

    4.4.2.1.4   avoid sudden braking

    4.4.2.1.5   avoid coarse/harsh steering inputs

    4.4.2.1.6   observe road conditions of low friction (e.g. ice, rain, diesel spills etc.)

4.4.2.2 avoid control inputs which unbalance the vehicle during dynamic situations

Plan 4.4.2.2 - do 1 AND 2 AND 3 AND 4 AND 5

    4.4.2.2.1   avoid sudden/severe braking

    4.4.2.2.2   avoid unprogressive/sudden throttle inputs

    4.4.2.2.3   generally avoid lifting off the accelerator whilst cornering

    4.4.2.2.4   generally avoid braking mid corner

    4.4.2.2.5   avoid heavy acceleration mid cornering

    4.4.2.2.6   generally avoid sudden or jerky control inputs

4.4.2.3 detect presence of a skid

Plan 4.4.2.3 - do 1 AND/OR 2 AND/OR 3. IF detection occurring during cornering THEN 4 IF detection occurring during braking THEN 5

    4.4.2.3.1   check visually any discrepancy between desired and actual vehicle speed/trajectory

    4.4.2.3.2   check via auditory modality screeching or 'scrubbing' of tyres on road surface

    4.4.2.3.3   check via haptic sensations the scrubbing of tyres on road surface through steering wheel

    4.4.2.3.4   check via haptic sensations the rapid decrease in steering wheel aligning torque

    4.4.2.3.5   check via proprioception the 'bump' felt through the vehicles seat as the road wheel(s) lock

4.4.2.4 correct skid

Plan 4.4.2.4 - IF car wheel spinning THEN 1. IF vehicle wheels lock under braking THEN 2. IF car understeering

THEN 3. IF vehicle oversteering THEN 4. IF vehicle in four-wheel slide arising due to an excessive braking manoeuvre THEN 5 ELSE 7

4.4.2.4.1 correct wheel spin

Plan 4.4.2.4.1 - do 1 THEN 2. IF wheel spin occurs when setting off from stationary position THEN 3 THEN 4 AND 2. IF engine note changes THEN 5 AND 6 THEN 7 AND 8. IF engine falters THEN repeat at step 3

    4.4.2.4.1.1 release pressure on accelerator (swiftly)

    4.4.2.4.1.2 reapply pressure on accelerator (smoothly)

    4.4.2.4.1.3 depress clutch (briskly)

    4.4.2.4.1.4 raise clutch

    4.4.2.4.1.5 hold clutch in position

    4.4.2.4.1.6 increase pressure on accelerator pedal

    4.4.2.4.1.7 raise clutch pedal fully (smoothly and gradually)

    4.4.2.4.1.8 depress accelerator an amount proportional to increasing speed without wheel spin

4.4.2.4.2 perform cadence braking

Plan 4.4.2.4.2 - IF locking/skidding wheel detected THEN 1 THEN 2 IF wheels lock again THEN repeat plan. IF performing an emergency braking manoeuvre THEN 3 before repeating plan

    4.4.2.4.2.1 release pressure on brake pedal (rapidly)

    4.4.2.4.2.2 reapply pressure on brake pedal (rapidly but smoothly)

    4.4.2.4.2.3 allow wheels to lock momentarily

4.4.2.4.3 deal with understeer

Plan 4.4.2.4.3 - do 1 THEN 2 IF step 1 removes understeer situation THEN exit

    4.4.2.4.3.1 remove the cause of understeer

    Plan 4.4.2.4.3.1 - do 1 OR 1 AND 2

        4.4.2.4.3.1.1 release the accelerator

        4.4.2.4.3.1.2 depress the clutch pedal fully (quickly)

    4.4.2.4.3.2 correct the understeer condition

    Plan 4.4.2.4.3.2 - IF circumstances permit THEN 2 IF understeer ceases THEN 3 AND 4 ELSE 1 THEN 4

4.4.2.4.3.2.1 steer (vigorously) to attempt to regain original course

4.4.2.4.3.2.2 steer into the direction of the skid

4.4.2.4.3.2.3 steer the vehicle back onto course

4.4.2.4.3.2.4 apply power (gently)

Plan 4.4.2.4.3.2.4 - IF clutch depressed when removing the cause of understeer THEN 1. IF engine note changes THEN 2 AND 3 THEN 4 AND 5 ELSE do 5

4.4.2.4.3.2.4.1 raise clutch

4.4.2.4.3.2.4.2 hold clutch in position

4.4.2.4.3.2.4.3 increase pressure on accelerator pedal

4.4.2.4.3.2.4.4 raise clutch pedal fully (smoothly and gradually)

4.4.2.4.3.2.4.5 depress accelerator an amount proportional to desired rate of acceleration

4.4.2.4.4 deal with oversteer

Plan 4.4.2.4.4 - do 1 IF oversteer situation ceases THEN exit ELSE 2

4.4.2.4.4.1 << GO TO subroutine 5.4.2.4.3.1 'remove the cause of understeer' >>

4.4.2.4.4.2 correct oversteer

Plan 4.4.2.4.4.2 - WHILE 3 do 1 IF oversteer ceases THEN 2 AND 4

4.4.2.4.4.2.1 steer in the direction of the skid

4.4.2.4.4.2.2 steer vehicle back onto course (gently)

4.4.2.4.4.2.3 avoid steering excessively into direction of skid

4.4.2.4.4.2.4 apply power (gently)

4.4.2.4.5 deal with four-wheel skid arising due to an extreme braking manoeuvre

Plan 4.4.2.4.5 - do 1 THEN 2 AND 3. IF four-wheel slide ceases THEN 4

4.4.2.4.5.1 release the brake

4.4.2.4.5.2 << GO TO subroutine 4.4.2.4.3.1 'remove the cause of understeer' >>

4.4.2.4.5.3 steer car in desired direction

4.4.2.4.5.4 apply power (gently)

4.4.2.4.6 perform general skid correction tasks

Plan 4.4.2.4.6 - do 1 AND 5 (if required). IF 5 required AND the road is slippery THEN 4. IF 5 required AND road not slippery THEN 3 until steering control is regained. IF 4 not required THEN 2

4.4.2.4.6.1 remove cause of skid

Plan 4.4.2.4.6.1 - do 1 AND/OR 2

    4.4.2.4.6.1.1 release pressure on accelerator

    4.4.2.4.6.1.2 depress clutch

4.4.2.4.6.2 maintain pressure on the brake pedal

4.4.2.4.6.3 release the brakes

4.4.2.4.6.4 use cadence braking << GO TO subroutine 4.4.2.4.2 'cadence braking' >>

4.4.2.4.6.5 use steering to avoid collision

4.4.2.5 << GO TO subroutine 4.4.1.6 'collision mitigation' >>

5: PERFORM STRATEGIC DRIVING TASKS

Plan 5 - WHILE 7, do 1 AND 2 AND 3 AND 4 AND 5 AND 6

5.1 perform surveillance

Plan 5.1 - do 1 AND 2 AND 3 AND 4 AND 5

5.1.1 perform visual surveillance

Plan 5.1.1 - do 1 AND 2 (periodically). IF at point immediately prior to initiating a manoeuvre THEN 3

5.1.1.1 general forward visual surveillance

Plan 5.1.1.1 - WHILE 4 do 1 AND 2 AND 5

    5.1.1.1.1 continuously scan surroundings, shifting gaze frequently

    5.1.1.1.2 look well ahead

    5.1.1.1.3 adjust focal distance relative to speed and road location

Plan 5.1.1.1.3 - IF main road driving THEN 1. IF urban driving THEN 2. IF rural driving THEN 3

    5.1.1.1.3.1 focus at further distances

    5.1.1.1.3.2 view road ahead to next junction

    5.1.1.1.3.3 in rural areas view road layout/environment well ahead

       5.1.1.1.4 avoid fixating gaze on road immediately ahead

       5.1.1.1.5 watch for hazards related to the road surface (pot-holes, oil spills etc.)

    5.1.1.2 perform rearward surveillance

Plan 5.1.1.2 - do 1 AND/OR 2 AND/OR 3 and repeat in whatever combination increases view of road

      5.1.1.2.1 glance/look in rear-view mirror

      5.1.1.2.2 glance/look in offside wing mirror

      5.1.1.2.3 glance/look in nearside wing mirror

    5.1.1.3 check blind spots

Plan 5.1.1.3 - do 1 AND/OR 2 and repeat in whatever combination increases view of road

      5.1.1.3.1 glance over right shoulder

      5.1.1.3.2 glance over left shoulder

5.1.2 perform auditory surveillance

Plan 5.1.2 - do 1 AND 2 AND 3

    5.1.2.1 monitor vehicle sounds

    Plan 5.1.2.1 - do 1 AND 2 AND 3

      5.1.2.1.1 monitor engine/transmission/exhaust sounds

      5.1.2.1.2 monitor tyre sounds

      5.1.2.1.3 monitor other vehicle related sounds

    5.1.2.2 monitor environmental/other sounds

    Plan 5.1.2.2 - do 1 AND 2 AND 3

      5.1.2.2.1 monitor sounds emitted by other vehicles

      5.1.2.2.2 monitor sounds emitted by other road related events

      5.1.2.2.3 monitor sounds linked to relevant non-road related events

    5.1.2.3 try to identify source of unusual sounds

    Plan 5.1.2.3 - do 1 AND/OR 2 AND/OR 3 AND/OR 4

      5.1.2.3.1 look in direction of noise source

      5.1.2.3.2 open window to improve audibility

      5.1.2.3.3 note whether noise is continuous or intermittent

      5.1.2.3.4 note whether intensity is increasing or decreasing (as indication that vehicle is passing noise source)

5.1.3 perform olfactory surveillance

Plan 5.1.3 - do 1 AND 2

    5.1.3.1 check for indications of external origin

    Plan 5.1.3.1 - do in any order

5.1.3.1.1 check for smoke/steam from vehicle ahead's exhaust

5.1.3.1.2 check immediate external environment for any indications of source

5.1.3.2 check for indications of internal origin

Plan 5.1.3.2 - do in any order

5.1.3.2.1 check for smoke from dashboard (indicating electrical fault)

5.1.3.2.2 check handbrake is not still applied

5.1.3.2.3 check engine temperature gauge

5.1.4 observe behaviour of other drivers

Plan 5.1.4 - do 1 AND 2 AND 3 AND 4 AND 5 AND 6 AND 7

5.1.4.1 note drivers who frequently change lane

5.1.4.2 note drivers who frequently change speed

5.1.4.3 note drivers who neglect to signal

5.1.4.4 note drivers who brake suddenly

5.1.4.5 note unconfident/unsure drivers

5.1.4.6 note aggressive drivers

5.1.4.7 note inattentive drivers

5.1.5 perform surveillance of own vehicle

Plan 5.1.5 - do 1 AND 2 AND 3

5.1.5.1 check instrument panel displays (regularly) to keep abreast of vehicle's operating characteristics

Plan 5.1.5.1 - do 1 frequently AND 2 WHEN accelerating ELSE periodically AND 3 periodically AND 4 when required

5.1.5.1.1 observe speedometer

Plan 5.1.5.1.1 - WHILE 1 do 2,3,4 as required

5.1.5.1.1.1 observe speedometer periodically

5.1.5.1.1.2 check speed whenever there is a change in the legal limit

5.1.5.1.1.3 check speed frequently after sustained high speeds

5.1.5.1.1.4 pay particular attention to speed in urban areas

5.1.5.1.2 observe rev counter

5.1.5.1.3 observe fuel gauge periodically

5.1.5.1.3 monitor engine temperature gauge

5.1.5.1.4 monitor warning lights

5.1.5.2 note any unusual performance in the car's operation
<< GO TO subroutine 5.1 'surveillance' >>

5.1.5.3 react to anything within the cabin that would adversely affect driving performance

5.2 perform navigation

Plan 5.2 - IF route already travelled previously THEN 1, ELSE 2 AND/OR 3 OR 4

5.2.1 use previous/local knowledge to maximum advantage

5.2.2 plan route in advance

5.2.3 use road atlas/street map

5.2.4 follow instructions from knowledgeable passenger

5.3 comply with rules

Plan 5.3 - do 1 AND 2

5.3.1 act on advice/instructions/rules/guidance provided by the Highway Code

5.3.2 respond to directions/instructions from police/authorized persons

5.4 respond to environmental conditions

Plan 5.4 - do 1 AND/OR 2

5.4.1 respond to weather conditions

Plan 5.4.1 - WHILE 1 AND 7 do 2 AND/OR 3 AND/OR 4 AND/OR 5 AND/OR 6 as required for the particular weather conditions

5.4.1.1 deal with limited visibility

Plan 5.4.1.1 - WHILE 1 do 2 AND/OR 3 AND/OR 4 as required

5.4.1.1.1 perform general adjustments to driving

Plan 5.4.1.1.1 - do 1 AND 2 AND 3 AND 4

5.4.1.1.1.1 drive more slowly than under normal conditions

5.4.1.1.1.2 increase following distance to compensate for decreased visibility

5.4.1.1.1.3 drive in lane that permits greater separation from oncoming traffic

5.4.1.1.1.4 increase degree of attentiveness

5.4.1.1.2 deal with limited visibility through windscreen

Plan 5.4.1.1.2 - do 1 AND 2 as required

5.4.1.1.2.1 deal with poor visibility due to rain

Plan 5.4.1.1.2.1 - do 1 AND 2 IF rain light AND windscreen dirty/greasy THEN 3

5.4.1.1.2.1.1 turn on windscreen wipers

5.4.1.1.2.1.2 select appropriate wiper speed

Plan 5.4.1.1.2.1.2 - do 1 AND 2

5.4.1.1.2.1.2.1 ensure sweep of wipers sufficiently clears screen of water

5.4.1.1.2.1.2.2 avoid causing the wipers to screech/chatter across screen

5.4.1.1.2.1.3 operate windscreen washers

5.4.1.1.2.2 deal with poor visibility due to condensation

Plan 5.4.1.1.2.2 - do 1 THEN 2 THEN 3. IF demister slow to clear condensation THEN 4 ELSE 5

5.4.1.1.2.2.1 select demist on the vehicles interior heater controls

5.4.1.1.2.2.2 turn on interior heater fan

5.4.1.1.2.2.3 adjust controls to increase demisting performance

5.4.1.1.2.2.4 open window slightly

5.4.1.1.2.2.5 remove heavy moisture with suitable cloth

5.4.1.1.3 deal with limited visibility through rear screen

Plan 5.4.1.1.3 - IF condensation on inside of rear screen THEN 1. IF road spray on outside of screen THEN 2. IF rear screen dirty OR vision otherwise restricted THEN 2 AND 3

5.4.1.1.3.1 operate rear demister

5.4.1.1.3.2 operate rear wiper

5.4.1.1.3.3 operate rear screen washer

5.4.1.1.4 deal with limited visibility through side windows

Plan 5.4.1.1.4 - IF windows misted on the inside THEN 1 AND 2 (to speed up demisting) ELSE 3. IF windows misted up on the outside THEN 4 ELSE 3

5.4.1.1.4.1 operate interior heater

5.4.1.1.4.2 open window slightly

5.4.1.1.4.3 use cloth to clear window

5.4.1.1.4.4 wind window right down to the bottom then right up to the top

5.4.1.2 deal with rain or fog

Plan 5.4.1.2 - WHILE 1 do 2 as required

    5.4.1.2.1 deal with limited visibility due to rain or fog

    Plan 5.4.1.2.1 - WHILE 1 do 2 AND 3 as required. IF fog severe AND following vehicle cannot be seen THEN 4. IF rain/fog very severe AND restricting vision to a dangerous extent THEN 5

        5.4.1.2.1.1 reduce speed so as not to overdrive visibility

        5.4.1.2.1.2 use road markings and other vehicle lights as additional longitudinal and lateral cues

        5.4.1.2.1.3 turn on main beam headlights

        5.4.1.2.1.4 use high intensity rear fog lights

        5.4.1.2.1.5 stop at roadside to wait out severe downpours/ extreme fog

    5.4.1.2.2 deal with wet roads

    Plan 5.4.1.2.2 - do 1 AND 2 AND 3 THEN 4 as required

        5.4.1.2.2.1 employ at least double braking distances

        5.4.1.2.2.2 exercise particular care if wet roads follow dry spell

        5.4.1.2.2.3 generally avoid large puddles/standing water

        5.4.1.2.2.4 deal with aquaplaning

        Plan 5.4.1.2.2.4 - WHILE 1 do 2

            5.4.1.2.2.4.1 attempt to maintain straight course

            5.4.1.2.2.4.2 << GO TO subroutine 4.4.2 'skid control' >>

    5.4.1.3 deal with glare from the sun

    Plan 5.4.1.3 - do 1 AND/OR 2 AND/OR 3

        5.4.1.3.1 adjust sun visors to shield eyes without obstructing view

        5.4.1.3.2 wear sunglasses

        5.4.1.3.3 look down at roadway in front of car (not directly into sun)

    5.4.1.4 deal with extreme temperatures

    Plan 5.4.1.4 - do 1 OR 2

        5.4.1.4.1 deal with conditions of extreme heat

        Plan 5.4.1.4.1 - do 1 IF engine overheating THEN 2

            5.4.1.4.1.1 watch temperature gauge for signs of overheating

            Plan 5.4.1.4.1.1 - do 1 AND 2 AND 3

5.4.1.4.1.1.1 observe current temperature

5.4.1.4.1.1.2 observe rate of engine temperature change

5.4.1.4.1.1.3 avoid putting the engine under high load

Plan 5.4.1.4.1.1.3 - do 1 AND 2

    5.4.1.4.1.1.3.1 use high gears

    5.4.1.4.1.1.3.2 use light throttle openings

5.4.1.4.1.2 deal with overheating engine

Plan 5.4.1.4.1.2 - do 1 THEN 2 THEN 3 IF engine still overheating AND waiting in traffic THEN 4 IF engine still overheating THEN 5 THEN 6 THEN 7

    5.4.1.4.1.2.1 drive in high gear

    5.4.1.4.1.2.2 open windows

    5.4.1.4.1.2.3 run interior heater in hottest position

    5.4.1.4.1.2.4 run engine on slightly faster idle

    5.4.1.4.1.2.5 pull over and stop

    5.4.1.4.1.2.6 open bonnet

    5.4.1.4.1.2.7 run engine for a few minutes before turning off ignition

5.4.1.4.2 deal with conditions of extreme cold

Plan 5.4.1.4.2 - WHILE 1 IF engine at normal operating temperature THEN 2 do 3 AND 4 WHILE 5 do 6 (be prepared to perform step 6 much more than is normal for everyday driving)

    5.4.1.4.2.1 observe for icy/slippery patches on roadway

    5.4.1.4.2.2 turn on interior heater as required

    5.4.1.4.2.3 exercise extreme caution when braking

    5.4.1.4.2.4 exercise extreme caution when cornering

    5.4.1.4.2.5 deal with snow conditions

    Plan 5.4.1.4.2.5 - WHILE 1 do 2 AND 3 as required AND 4 (speed perception suffers in snow/ conditions of poor visibility). WHILE 5 do 6 AND 7 as a matter of course IF road conditions are particularly slippery.

        5.4.1.4.2.5.1 avoid using full beam (snow reflects light back into drivers' eyes)

        5.4.1.4.2.5.2 use main beam

5.4.1.4.2.5.3 use wipers

5.4.1.4.2.5.4 observe speedometer

5.4.1.4.2.5.5 (where possible) position vehicle in tyre tracks straddling centre mound of uncompacted snow

5.4.1.4.2.5.6 << GO TO subroutine 4.4.2.4.2 'cadence braking' >>

5.4.1.4.2.5.7 reduce engine torque to drive wheels

Plan 5.4.1.4.2.5.7 - WHILE 1 IF setting off from standstill THEN 2 ELSE 3

5.4.1.4.2.5.7.1 avoid large/sudden throttle openings

5.4.1.4.2.5.7.2 pull away using 2nd gear

5.4.1.4.2.5.7.3 use higher gears than would ordinarily be the case

5.4.1.4.2.6 << GO TO subroutine 4.4.2 'skid control' >>

5.4.1.5 deal with windy conditions

Plan 5.4.1.5 - WHILE 1 do 2

5.4.1.5.1 drive at lower than normal speed

5.4.1.5.2 deal with vehicle's tendency to 'roll steer' in response to wind gusts

Plan 5.4.1.5.2 - WHILE 1 do 2 AND 3 AND 4

5.4.1.5.2.1 grasp steering wheel firmly

5.4.1.5.2.2 steer toward wind when car's lateral positioning is altered by wind force

5.4.1.5.2.3 avoid oversteering in reacting to gusts

5.4.1.5.2.4 anticipate need for steering corrections when wind is screened by hills/buildings/larger vehicles

5.4.1.6 observe for 'micro climates'

Plan 5.4.1.6 - do 1 AND 2 AND 3 AND 4 AND 5

5.4.1.6.1 observe for valley bottoms (where pockets of fog/ ice may linger)

5.4.1.6.2 observe for shaded hillsides/slopes

5.4.1.6.3 observe for large areas of shadow cast by trees

5.4.1.6.4 observe for patchy fog

5.4.1.6.5 observe road conditions on bridges (here road is cooled on all sides and may be icy when surrounding roads are not)

5.4.1.7 avoid inappropriate driving for the conditions

Plan 5.4.1.7 - do 1 THEN 2 AND 3

    5.4.1.7.1 consider how much grip the vehicle is likely to possess in the current environmental conditions

    5.4.1.7.2 issue steering/throttle/braking inputs appropriate for current environmental conditions

    5.4.1.7.3 drive at a speed in which the available visibility provides adequate stopping distance

5.4.2 drive at night

Plan 5.4.2 - WHILE 1 IF driving in urban situation THEN 2. IF driving in rural situation THEN 3. IF driving at dusk THEN 4.

    5.4.2.1 perform general night driving tasks

Plan 5.4.2.1 - do 1 AND 2 AND 3 AND 4 AND 5 AND 6 AND 7 AND 8 AND 9

    5.4.2.1.1 drive with main beam headlights

    5.4.2.1.2 adopt appropriate speed for night driving

    Plan 5.4.2.1.2 - do 1 AND 2

        5.4.2.1.2.1 drive more slowly than under similar circumstances during daylight

        5.4.2.1.2.2 maintain speed that permits stopping within distance illuminated by headlights

    5.4.2.1.3 watch for dark or dim objects on roadway

    5.4.2.1.4 watch beyond headlight beams (for slow moving/ unlit vehicles/curves/road obstructions/ defects/ pedestrians/ animals)

    5.4.2.1.5 allow greater margin of safety in performance of manoeuvres than during daylight

    Plan 5.4.2.1.5 - do 1 AND 2

        5.4.2.1.5.1 increase following distances

        5.4.2.1.5.2 increase distance and time for an acceptable passing opportunity

        5.4.2.1.6 use headlight beams of other vehicles as indication of direction of approaching vehicles

    5.4.2.1.7 use cat's eyes/reflective signs/markers to gauge direction of road and presence of hazards

    5.4.2.1.8 keep car well ventilated

5.4.2.1.9 stop every (approximately) 2 hours when driving for an extended period

5.4.2.2 undertake urban night driving

Plan 5.4.2.2 - WHILE 1 do 2 AND 3

5.4.2.2.1 do not use high beam

5.4.2.2.2 check headlights are on (as ambient lighting makes it easy to forget)

5.4.2.2.3 watch for pedestrians/unlit vehicles/objects on the road/curbside

5.4.2.3 undertake rural night driving

Plan 5.4.2.3 - WHILE 1 do 2 AND 3. IF lights of following vehicle dazzling THEN 4. do 5 IF hazards detected AND evasive action required to avoid collision OR hazard THEN 6

5.4.2.3.1 generally use high beam headlights

Plan 5.4.2.3.1 - IF following vehicle THEN 1. IF vehicle oncoming THEN 1 AND 2. IF lights of oncoming vehicle especially bright THEN 3 AND/OR 4

5.4.2.3.1.1 maintain headlights on low beam

5.4.2.3.1.2 avoid looking directly at approaching vehicles' headlights

5.4.2.3.1.3 focus eyes to left side of road beyond oncoming vehicle

5.4.2.3.1.4 close one eye as vehicle draws near to save it until vehicle passes

5.4.2.3.2 use tail lights of lead vehicle to gauge closing rate

5.4.2.3.3 maintain safe following distance

5.4.2.3.4 flick rear-view mirror to night position

5.4.2.3.5 watch for pedestrians/animals/unlit vehicles on or beside road

5.4.2.3.6 << GO TO subroutine 4.4.1 'take evasive action' >>

5.4.2.4 drive at dusk/dawn/dark days

Plan 5.4.2.4 - WHILE 1 do 2 IF wearing sunglasses THEN 3 IF sufficiently dark THEN 4

5.4.2.4.1 drive slower giving increased attention to traffic

5.4.2.4.2 use side lights

5.4.2.4.3 remove sunglasses

5.4.2.4.4 use main beam

5.5 perform (IAM) system of car control

Plan 5.5 - WHILE 1 AND 2 IF hazard necessitates a change in road position THEN 3 IF hazard necessitates a change in vehicle speed THEN 4 IF change in speed OR anticipated need for acceleration requires a future change of speed THEN 4 IF speed change OR anticipated speed change requires it THEN 5 IF step 1 requires it THEN overlap steps 4 AND 5 WHILE leaving/exiting hazard do 6

    5.5.1 use the system flexibly

    5.5.2 perform the information phase

Plan 5.5.2 - do 1 AND 2 AND 3 as required by road/traffic situation and nature of hazard/potential hazard

        5.5.2.1 take information from driving environment

        5.5.2.2 use information (hazard detection/anticipate)

        5.5.2.3 give information (to other road users)

    5.5.3 perform the position phase

Plan 5.5.3 - WHILE 1 do 2

        5.5.3.1 take account of other road users

        5.5.3.2 adopt position that permits hazards to be passed/dealt with safely (smoothly)

    5.5.4 perform the speed phase

Plan 5.5.4 - WHILE 1 do 2

        5.5.4.1 make good use of acceleration sense (perception of vehicle speed/relative speeds)

        5.5.4.2 adjust speed in order to complete manoeuvre safely (smoothly)

    5.5.5 perform the gear phase

Plan 5.5.5 - WHILE 3 AND 4 do 1 IF nature of hazard/ manoeuvre dictates THEN 2

        5.5.5.1 engage correct gear for speed that has been selected in order to negotiate hazard safely (smoothly)

        5.5.5.2 make the gear change before braking

        5.5.5.3 (generally) avoid late braking

        5.5.5.4 (generally) avoid snatched gear changes

    5.5.6 perform the acceleration phase

Plan 5.5.6 - WHILE 1 IF conditions permit THEN 2

        5.5.6.1 take account of road and traffic conditions ahead

      Plan 5.5.6.1 - do 1 AND 2

           5.5.6.1.1 take account of current speed

5.5.6.1.2 take account of speed of other road users

5.5.6.2 accelerate safely (smoothly) away from hazard

5.6 exhibit vehicle/mechanical sympathy

Plan 5.6 - do 1 AND 2 AND 3 AND 4 AND 5

5.6.1 (generally) avoid putting vehicles engine under undue stress

Plan 5.6.1 - do 1 AND 2 OR 3 AND 4 AND 5

5.6.1.1 avoid sustained full throttle operation

5.6.1.2 avoid sustained running at engine speeds above 5750RPM

5.6.1.3 avoid sustained running at engine speeds as advised by vehicle owner's manual

5.6.1.4 avoid high load operation at engine speeds >2000RPM

5.6.1.5 avoid abrupt accelerator activation

5.6.2 (generally) avoid putting vehicle's clutch under undue stress

Plan 5.6.2 - WHILE 4 do 1 AND 2 AND 3 (except when required for full power acceleration from standstill)

5.6.2.1 never hold vehicle on gradient by slipping clutch

5.6.2.2 avoid slipping clutch for longer than 15 seconds

5.6.2.3 generally avoid slipping clutch at engine speeds <4000RPM

5.6.2.4 engage clutch briskly (smoothly)

5.6.3 (generally) avoid putting vehicles transmission under undue stress

Plan 5.6.3 - WHILE 1 do 2 AND 3 AND 4 AND 5 AND 6 AND 7 IF engine speed is vastly different from road speed THEN WHILE 4 do 8 ELSE 9 THEN 8

5.6.3.1 use feel through gear lever to guide lever into desired gear position

5.6.3.2 always fully depress clutch during gear changes

5.6.3.3 avoid clutchless 'racing' changes

5.6.3.4 avoid forcing gear lever into desired selector gate

5.6.3.5 never engage reverse whilst vehicle is in motion

5.6.3.6 avoid block changes when engine speed is vastly different from road speed

5.6.3.7 avoid wheel spin

5.6.3.8 use double-declutch procedure

5.6.3.9 brake to target speed

5.6.4 (generally) avoid putting the vehicles suspension system under undue stress

Plan 5.6.4 - do 1 AND 2 AND 3

5.6.4.1 avoid excessive tyre wear

Plan 5.6.4.1 - do 1 AND 2 AND 3

5.6.4.1.1 do not overload vehicle

5.6.4.1.2 do not steer vehicle whilst completely stationary

5.6.4.1.3 generally avoid frequent high load vehicle dynamics

Plan 5.6.4.1.3 - do 1 AND/OR 2 AND/OR 3

5.6.4.1.3.1 avoid frequent heavy braking

5.6.4.1.3.2 avoid frequent hard acceleration

5.6.4.1.3.3 avoid frequent hard cornering

5.6.4.2 avoid excessive suspension wear

Plan 5.6.4.2 - do 1 AND 2 AND 3

5.6.4.2.1 avoid sustained running near edge of road

5.6.4.2.2 avoid running over rough terrain at high speeds

5.6.4.2.3 avoid excessive loading of vehicle

5.6.4.3 avoid suspension damage

Plan 5.6.4.3 - do 1 AND 2 AND 3

5.6.4.3.1 never run wheels up kerbs

5.6.4.3.2 avoid potholes/other road damage

5.6.4.3.3 never allow the vehicles suspension to 'bottom out'

5.6.5 (generally) avoid putting the vehicle's brakes under undue stress

Plan 5.6.5 - do 1 AND 2 AND 3 AND 4 THEN 5 WHEN safe to do so

5.6.5.1 avoid bringing the brakes to the point of 'brake fade'

5.6.5.2 never use the handbrake whilst vehicle is in motion

5.6.5.3 avoid sustained application of brakes for <2/3 minutes

5.6.5.4 (to prevent warped brake discs) release brake pedal after sustained/heavy braking manoeuvre

5.6.5.5 use handbrake to hold vehicle stationary after coming to a halt

5.7 exhibit appropriate driver attitude/deportment

Plan 5.7 - do 1 AND 2 WHILE 3,4,5
  5.7.1 exhibit good general skill characteristics
  Plan 5.7.1 - do 1 AND 2 AND 3 AND 4 AND 5 AND 6
          AND 7
    5.7.1.1 maintain good level of attention at all times
    5.7.1.2 perform accurate observation
    5.7.1.3 match vehicle speed/trajectory to the situation
    5.7.1.4 maintain high levels of awareness of risks inherent in
              particular road/traffic situation
    5.7.1.5 anticipate risks/(potential) hazards
    5.7.1.6 act in a manner appropriate to minimising identified
              risks
    5.7.1.7 skilful use of controls
  5.7.2 exhibit favourable general attitudinal characteristics
  Plan 5.7.2 - do 1 AND 2 AND 3 AND 4 AND 5
    5.7.2.1 maintain favourable attitude towards other road users
    Plan 5.7.2.1 - do 1 AND 2 AND 3 AND 4 AND 5
      5.7.2.1.1 avoid selfish behaviour
      5.7.2.1.2 avoid aggressive behaviour
      5.7.2.1.3 ensure considerate/constructive approach to other
                road users
      5.7.2.1.4 ensure tolerance at all times
      5.7.2.1.5 maintain sense of responsibility for other's safety
    5.7.2.2 remain patient at all times
    5.7.2.3 remain critically self-aware of driver limitations
    5.7.2.4 remain aware of vehicle limitations
    5.7.2.5 maintain sense of responsibility for own safety
  5.7.3 avoid 'red mist'
  5.7.4 concentrate on the driving task
  5.7.5 remain relaxed and unflustered
  5.7.6 drive with confidence

6: PERFORM POST-DRIVE TASKS

Plan 6 – do in order
6.1 park the vehicle
Plan 6.1 - initiate 1 OR 2 OR 3 OR 4 as required
  6.1.1 reverse parking
  6.1.2 parallel parking

6.1.3 forward parking

6.1.4 parking in garages

6.2 make the vehicle safe

Plan 6.2 - do 1 THEN 2 THEN 3. IF car on slope THEN 4 THEN 5 THEN 6 THEN 7 THEN 8

6.2.1 bring the vehicle to complete halt with the footbrake

6.2.2 apply the handbrake << GO TO subroutine 2.4.4.2 'apply handbrake' >>

6.2.3 release footbrake

6.2.4 turn front wheels in towards the kerb

6.2.5 turn ignition key to position 3

6.2.6 remove key from ignition

6.2.7 engage steering lock

6.2.8 remove seat belt

6.3 leave the vehicle

Plan 6.3 - do 1 AND 2. IF unsafe to open door THEN wait until it is safe ELSE 3 THEN 4 THEN 5 THEN 6 THEN 7 THEN 8 THEN 9 THEN 10

6.3.1 turn off electrical systems

6.3.2 check that it is safe to open door

6.3.3 operate interior door handle

6.3.4 push door open

6.3.5 swing legs out of footwell onto road

6.3.6 alight from vehicle

6.3.7 shut door

6.3.8 lock door

6.3.9 ensure all other doors are locked

6.3.10 walk away

SA Requirements Capture Table – Individual SA

TASK/GOAL NUMBER	TASK/GOAL DESCRIPTION	LEVEL 1 SA REQUIREMENTS	LEVEL 2 SA REQUIREMENTS	LEVEL 3 SA REQUIREMENTS
		Prompts: *What does the driver need to perceive in order to complete this task/goal successfully?*	*Prompts:* *What does the driver need to comprehend in order to complete this task/goal successfully?* *How is the perceived information combined or integrated?*	*Prompts:* *In what way is the situation likely to change in the near future?* *What is likely to happen next?* *How could perceiving or comprehending features of the current situation help the driver to perform in these future situations?*

Extend the blank rows as required…

SA Requirements Capture Table – Team/Compatible SA

TASK/GOAL NUMBER	TASK/GOAL DESCRIPTION	INFORMATION ELEMENTS	OWNERSHIP	COMMUNICATION	COMPATIBILITY
		Prompts: What discrete pieces of information does the driver need to have knowledge of in order to perform this task successfully?	Prompts: Who owns this information? Who produces this information? Who/what displays this information? Who/what has a unique view of this information?	Prompts: Are information elements 'owned' by more than one part of the system (human and/or technical)? If 'no' leave cell blank. If 'yes' note how that information is communicated by one system agent to another.	Prompts: In cases where information elements have multiple owners, provide an assessment of how compatible the same element is (in terms of how it is currently presented/communicated to each system agent). Is it compatible in terms of speed of communication? Mode of communication? Meaning of communication?

Extend the blank rows as required. . . .

Further Reading

Harvey, C. and Stanton, N. A. (2013). *Usability Evaluation for In-Vehicle Systems*. Boca Raton: CRC Press.

Salmon, P. M., Stanton, N. A., Walker, G. H., and Jenkins, D. P. (2009). *Distributed Situation Awareness: Advances in Theory, Measurement and Application to Teamwork*. Farnham: Ashgate.

Stanton, N. A., Salmon, P. M., Walker, G. H., Baber, C., and Jenkins, D. P. (2013). *Human Factors Methods: A Practical Guide for Engineering and Design*, 2nd Edition. Farnham: Ashgate.

Walker, G. H., Stanton, N. A., and Salmon, P. M. (2015). *Human Factors in Automotive Engineering and Design*. Farnham: Ashgate.

References

Ackerman, R. K. (1998). New display advances brighten situational awareness picture. *Combat Edge* [online]. Available from: www.highbeam.com/doc/1P3-33041082.html [Accessed 19 February 2013].

Ackerman, R. K. (2005). Army intelligence digitizes situational awareness. *Signal* [online]. Available from: http://sncorp.com/ [Accessed 19 February 2015].

Adams, M. J., Tenney, Y. J., and Pew, R. W. (1995). Situation awareness and the cognitive management of complex systems. *Human Factors, 37*(1), 85–104.

Akao, Y. (1990). *Quality Function Deployment: Integrating Customer Requirements into Product Design*. New York: Productivity Press.

Allen, R. E. (1984). *The Pocket Oxford Dictionary of Current English*. Oxford: Clarendon.

Allen, T. M., Lunenfeld, H., and Alexander, G. J. (1971). Driver information needs. *Highway Research Record, 366*, 102–115.

Anderson, J. (1983). *The Architecture of Cognition*. Cambridge, MA: Harvard University Press.

Annett, J. (2002). Target paper: Subjective rating scales: Science or art? *Ergonomics, 45*(14), 966–987.

Annett, J., Duncan, K. D., Stammers, K. B., and Gray, M. J. (1971). *Task Analysis*. London: Her Majesty's Stationery Office.

Annett, J. and Kay, H. (1957). Knowledge of results and skilled performance. *Occupational Psychology, 31*(2), 69–79.

Annett, J. and Stanton, N. A. (1998). Research and developments in task analysis. *Ergonomics, 41*(11), 1529–1536. Reprinted in: J. Annett & N. A. Stanton (2000, eds.) *Task Analysis*. London: Taylor and Francis, pp. 1–8.

Annett, J. and Stanton, N. A. (2000). Team work: A problem for Ergonomics? *Ergonomics, 43,* 1045–1051.

Artman, H. and Garbis, C. (1998). Team communication and coordination as distributed cognition. In T. Green, L. Bannon, C. Warren, and J. Buckley (Eds.) *Proceedings of 9th Conference of Cognitive Ergonomics: Cognition and Cooperation,* Limerick, Ireland, pp. 151–156.

Asvin, G. (2008). *Fleet Telematics: Real-Time Management and Planning of Commercial Vehicle Operations.* Boston: Springer.

Ausubel, D. (1963). *The Psychology of Meaningful Verbal Learning.* New York: Grune and Stratton.

Autocar (Jan. 2009). The Eurofighter effect. Available from: www.autocar.co.uk/opinion/tester-s-notes/eurofighter-effect [Accessed 15 February 2018].

Auto Express (2017). Infiniti Q50 review. Available from: www.autoexpress.co.uk/infiniti/q50 [Accessed 15 February 2018].

Baber, C. (2004). Personal communication.

Badham, R., Clegg, C.W., & Wall, T. (2000). Socio-technical theory. In W. Karwowski (Ed.), *Handbook of Ergonomics.* New York: Wiley.

Banbury, S. P., Croft, D. G., Macken, W. J., and Jones, D. M. (2004). A cognitive streaming account of situation awareness. In S. Banbury and S. Tremblay (Eds.) *A Cognitive Approach to Situation Awareness: Theory and Application.* Aldershot: Ashgate.

Barthorpe, F. (May 2000). Sound engineering: Good vibrations. *Automotive Engineer, 21,* 31–35.

Bartlett, F. C. (1932). *Remembering: A Study in Experimental and Social Psychology.* Cambridge: Cambridge University Press.

Bashford, G. D. (June/July 1978). Influence of king pin inclination on steering effort. *Automotive Engineer,* 49–50.

Baxter, G., Besnard, D., and Riley, D. (2007). Cognitive mismatches in the cockpit: Will they ever be a thing of the past? *Applied Ergonomics, 38*(4), 417–423.

Becker, K., Castro, P., Boyle, A., and Eichhorn, U. (October/November 1996). New launch: Ford is excited as a new baby, the Ka, is born. *Automotive Engineer,* 31–40.

Bedinger, M., Walker, G. H., Piecyk, M., and Greening, P. (2016). 21st century trucking: A trajectory for ergonomics and road freight. *Applied Ergonomics, 53,* 343–356.

Bedny, G. and Meister, D. (1999). Theory of activity and situation awareness. *International Journal of Cognitive Ergonomics, 3*(1), 63–72.

Bell, H. H. and Lyon, D. R. (2000). Using observer ratings to assess situation awareness. In M. R. Endsley (Ed.) *Situation Awareness Analysis and Measurement.* Mahwah: Lawrence Erlbaum Associates.

Billings, C. E. (1995). Situation awareness measurement and analysis: A commentary. *Proceedings of the International Conference on Experimental Analysis and Measurement of Situation Awareness,* Daytona Beach: Embry-Riddle Aeronautical University Press.

Blandford, A. and Wong, B. L. W. (2004). Situation awareness in emergency medical dispatch. *International Journal of Human-Computer Studies, 61*(4), 421–452.

Bleakley, A., Allard, J., and Hobbs, A. (2013). 'Achieving ensemble': Communication in orthopaedic surgical teams and the development of situation awareness—An observational study using live videotaped examples. *Advances in Health Sciences Education*, *18*(1), 33–56.

Bliss, J. P. and Acton, S. A. (2003). Alarm mistrust in automobiles: How collision alarm reliability affects driving. *Applied Ergonomics*, *34*(6), 499–509.

Bolia, R., Vidulich, M., Nelson, T., and Cook, M. (2007). A history lesson in the use of technology to support military decision making and command and control. In M. Cook, J. Noyes, and Y. Masakowski (Eds.), *Decision Making in Complex Environments* (pp. 191–200). Aldershot: Ashgate.

Bonsall, P. (1992). The influence of route guidance on route choice in urban networks. *Transportation*, *19*(1), 1–23.

Bonsall, P. W. and Palmer, I. (1999). Behavioural response to roadside variable message signs: Factors affecting compliance. In R. Emmerick and P. Nijkampt (Eds.) *Behavioural and Network Impacts of Driver Information Systems*. Farnham: Ashgate.

Bosworth, R., Trinick, J., Smith, T., and Horswill, S. (1996). Rover's system approach to achieving first class ride comfort for the new Rover 400. *Automotive Refinement. Selected Papers from Autotech 95*, ImechE/MEP, London, 7–9 November 1995.

Bourbousson, J., Poizat, G., Saury, J., and Seve, C. (2011). Description of dynamic shared knowledge: An exploratory study during a competitive team sports interaction. *Ergonomics*, *54*(2), 120138.

Bowen, J. T. (2012). A spatial analysis of FedEx and UPS: Hubs, spokes, and network structure. *Journal of Transport Geography*, *24*, 419–431.

Bratman, M. E. (1992). Shared cooperative activity. *The Philosophical Review*, *101*(2), 327–341.

Brindle, R. E. (1996). *Urban Road Classification and Local Street Function*. Vermont South: Australian Road Research Board.

Brookhuis, K. A., van Driel, C. J. G., Hof, T., van Arem, B., and Hoedemaeker, M. (2008). Driving with a congestion assistant: Mental workload and acceptance. *Applied Ergonomics*, *40*(6), 1019–1025.

Bryant, D. J., Lichacz, F. M. J., Hollands, J. G., and Baranski, J. V. (2004). Modelling situation awareness in an organisational context: Military command and control. In S. Banbury and S. Tremblay (Eds.) *A Cognitive Approach to Situation Awareness: Theory and Application*. Aldershot: Ashgate.

Carsten, O. and Venderhaegen, F. (2015). Situation awareness: Valid or fallacious? *Cognition, Technology & Work*, *17*(2): 157–158.

Cattell, R. B. (1966). The scree test for the number of factors. *Multivariate Behavioral Research*, *1*(2), 629–637.

Chase, W. G. and Simon, H. A. (1973). Perception in chess. *Cognitive Psychology*, *4*, 55–81.

Cherrett, T., Allen, J., McLeod, F., Maynard, S., Hickford, A., and Browne, M. (2012). Understanding urban freight activity–Key issues for freight planning. *Journal of Transport Geography*, *24*, 22–32.

Chorlton K. and Jamson, S. L. (2003). *Who Rides on Our Roads? An Exploratory Study of the UK Motorcycling Fleet*. Leeds: Institute for Transport Studies, University of Leeds.

Chorus, C., Molin, E. E., and Van Wee, B. (2006). Use and effects of Advanced Traveller Information Services (ATIS): A review of the literature. *Transport Reviews, 26*(2), 127–149.

Chowdhury, I. (2014). *A user-centred approach to road design: blending distributed situation awareness with self-explaining roads*. Doctoral thesis (unpublished), Heriot-Watt University.

Clarke, D. D., Ward, P., Bartle, C., and Truman, W. (2004). *In Depth Study of Motor-cycle Accidents: Research Report No. 54*. London: Department for Transport.

Clarke, D. D., Ward, P., Bartle, C., and Truman, W. (2007). The role of motorcyclist and other driver behaviour in two types of serious accident in the UK. *Accident Analysis & Prevention, 39*(5), 974–981.

Clegg, C. W. (2000). Sociotechnical principles for system design. *Applied Ergonomics, 31*(5), 463–477.

Coe, R. (September 2002). It's the effect size, stupid: What effect size is and why it is important. Paper presented at the *Annual Conference of the British Educational Research Association*, Exeter.

Cohen, J. (1988). *Statistical Power Analysis for the Behavioural Sciences*. Hillsdale, NJ: Lawrence Erlbaum Associates.

Cohen, J. (1990). Things I have learned (so far). *American Psychologist, 45*(12), 1304–1312.

Cohen, J. (December 1994). The earth is round ($p < 0.05$). *American Psychologist, 49*(12), 997–1003.

Collins, A. M. and Loftus, E. F. (1975). A spreading-activation theory of semantic processing. *Psychological Review, 82*(6), 407–428.

Collins, A. M. and Quillian, M. R. (1969). Retrieval time from semantic memory. *Journal of Verbal Learning and Verbal Behaviour, 8*(2), 240–248.

Collins, A. M. and Quillian, M. R. (1970). Does category size affect categorization time? *Journal of Verbal Learning and Verbal Behaviour, 9*(4), 432–438.

Connolly, T. and Aberg, L. (1993). Some contagion model of speeding. *Accident Analysis & Prevention, 25*(1), 57–66.

Coombs, M. (2010). *Conciliatory Modelling of Web Sources Using Automated Content Analysis: A Vignette on Drilling for Shale Gas in Upstate New York*. Worcester, NY: Diploma Media Research.

Craig, A. (1979). Nonparametric measures of sensory efficiency for sustained monitoring tasks. *Human Factors, 21*(1), 69–78.

Crandall, B., Klein, G., and Hoffman, R. (2006). *Working Minds: A Practitioner's Guide to Cognitive Task Analysis*. Cambridge, MA: MIT Press.

Cretchley, J., Rooney, D., and Gallois, C. (2010). Mapping a 40-year history with Leximancer: Themes and concepts in the Journal of Cross-Cultural Psychology. *Journal of Cross-Cultural Psychology, 41*(3), 318–328.

Daniels, S., Brijs, T., Nuyts, E., and Wets, G. (2010). Explaining variation in safety performance of roundabouts. *Accident Analysis & Prevention, 42*(2), 393–402.

Dekker, S. (2015). The danger of losing situation awareness. *Cognition, Technology & Work, 17*(2), 59–161.

Divey, S. T. (1991). *The Accident Liabilities of Advanced Drivers.* Crowthorne: Transport Road and Research Laboratory (TRRL).

Driskell, J. E. and Mullen, B. (2005). Social network analysis. In N. A. Stanton, A. Hedge, K. Brookhuis, E. Salas, and H. W. Hendrick (Eds.) *Handbook of Human Factors and Ergonomics Methods* (pp. 58.1–58.6). London: CRC.

Dul, J., Bruder, R., Buckle, P., Carayon, P., Falzon, P., Marras, W. S., Wilson, J. R., and van der Doelen, B. (2012). A strategy for human factors/ ergonomics: Developing the discipline and profession. *Ergonomics, 55*(4), 377–395.

Duncan, J., Williams, P., and Brown, I. D. (1991). Components of driving skill: Experience does not mean expertise. *Ergonomics, 34*(7), 919–937.

Dutton, G. (2011). Fleet management's magic box. *World Trade, 24*(2), 38–44.

Ellis, J. R. (1994). *Vehicle Handling Dynamics.* Chichester: Wiley-Blackwell.

Elvik, R. (2010). Why some road safety problems are more difficult to solve than others. *Accident Analysis & Prevention, 42*(4), 1089–1096.

Endsley, M. R. (1988). Situation awareness global assessment technique (SAGAT). *Proceedings of the National Aerospace and Electronics Conference (NAECON),* IEEE, New York, pp. 789–795.

Endsley, M. R. (1995). Toward a theory of situation awareness in dynamic systems. *Human Factors, 37*(1), 32–64.

Endsley, M. R. (2000). Direct measurement of situation awareness: validity and use of SAGAT. In M. R. Endsley, and D. G. Garland (Eds.), *Situation Awareness Analysis and Measurement* (pp. 147–173). Mahwah: Lawrence Erlbaum Associates.

Endsley, M. R. (2004). Situation awareness: Progress and directions. In S. Banbury & S. Tremblay (Eds.), *A Cognitive Approach to Situation Awareness: Theory, Measurement and Application* (pp. 317–341). Aldershot: Ashgate.

Endsley, M. R. (2015). Situation awareness misconceptions and misunderstandings. *Journal of Cognitive Engineering and Decision Making, 9*(1), 4–32.

Endsley, M. R., Bolte, B., and Jones, D. E. (2003). *Designing for Situation Awareness: An Approach to User-Centred Design.* London: Taylor and Francis.

Endsley, M. R. and Garland, D. J. (Eds.) (2000). *Situation Awareness Analysis and Measurement.* Mahwah: Lawrence Erlbaum Associates.

Endsley, M. R. and Jones, W. M. (2001). A model of inter- and intra-team situation awareness: Implications for design, training and measurement. In M. McNeese, E. Salas and M. Endsley (Eds.) *New Trends in Cooperative Activities: Understanding System Dynamics in Complex Environments,* Santa Monica: Human Factors and Ergonomics Society.

Endsley, M. R. and Robertson, M. M. (2000). Situation awareness in aircraft maintenance teams. *International Journal of Industrial Ergonomics, 26,* 301–325.

Ericsson, K. A. and Simon, H. A. (1993). *Protocol Analysis: Verbal Reports as Data*. Cambridge, MA: MIT Press.

Evans, L., and Herman, R. (1976). Note on driver adaptation to modified vehicle starting acceleration. *Human Factors, 18*(3), 235–240.

Eysenck, M. W. and Keane, M. T. (1990). *Cognitive Psychology*, 2nd edition. Hove, UK: Lawrence Erlbaum.

Farber, G. (1999, September). SAE safety and human factors standards for ITS: Driver access to navigation systems in moving vehicles. Paper presented at Latest Developments in Intelligent Car Safety Systems, Mayfair, London.

Faul, F. and Erdfelder, E. (1992). *GPOWER: A priori, post hoc, and compromise power analyses for MS-DOS* [Computer software]. Bonn: Bonn University, Department of Psychology.

Field, E. and Harris, D. (1998). A comparative survey of the utility of cross-cockpit linkages and autoflight systems' backfeed to the control inceptors of commercial aircraft. *Ergonomics, 41*(10), 1462–1477.

Fioratou, E., Flin, R., Glavin, R., and Patey, R. (2010). Beyond monitoring: Distributed situation awareness in anaesthesia. *British Journal Anaesthesia, 105*(1), 83–90.

Fiore, S. M., Ross, K. G., and Jentsch, F. (2012). A team cognitive readiness framework for small-unit training. *Journal of Cognitive Engineering and Decision Making 6*(3): 325–349.

Fiore, S. M. and Salas, E. (2004). Why we need team cognition. In E. Salas and S.M. Fiore (Eds.), *Team Cognition* (pp. 235–248). Washington, DC: American Psychological Association.

Flach, J. (1995). Situation awareness: Proceed with caution. *Human Factors: The Journal of the Human Factors and Ergonomics Society, 37*(1), 149–157.

Flach, J., Mulder, M., and Paassen, M. M. (2004). The concept of the situation in psychology. In S. Banbury and S. Tremblay (Eds.) *A Cognitive Approach to Situation Awareness: Theory and Application* (pp. 42–60). Aldershot: Ashgate.

Fracker, M. (1991). *Measures of situation awareness: Review and future directions*. Report No. AL-TR-1991–0128, Wright Patterson Air Force Base, OH: Armstrong Laboratories: Crew Systems Directorate.

Giannopoulos, G. A. (1996). Implications of European transport telematics an advanced logistics and distribution. *Transport Logistics, 1*(1), 31–49.

Gibson, J. J. (1977). The theory of affordances. In R. E. Shaw and J. Bransford (Eds.) *Perceiving, Acting and Knowing* (pp. 67–82). Hillsdale, NJ: Lawrence Erlbaum Associates.

Gibson, J. J. (1979). *The Ecological Approach to Visual Perception*. Boston: Houghton Mifflin.

Gillespie, T. D. and Segel, L. (1983). Influence of front-wheel drive on vehicle handling at low levels of lateral acceleration. *Road Vehicle Handling*, C114/83, 61–68.

Gkikas, N. (2012). Driving in the Era of IVIS and ADAS. In L. Dorn (Ed.) *Driver Behaviour and Training Volume V* (pp. 417–428). Aldershot: Ashgate.

Gobet, F. (1998). Expert memory: A comparison of four theories. *Cognition*, *66*(2), 115–152.

Godthelp, H. and Kappler, W. D. (1988). Effects of vehicle handling characteristics on driving strategy. *Human Factors*, *30*(2), 219–229.

Golightly, D., Ryan, B., Dadashi, N., Pickup, L., and Wilson, J. R. (2013). Use of scenarios and function analyses to understand the impact of situation awareness on safe and effective work on rail tracks. *Safety Science*, *56*, 52–62.

Golightly, D., Wilson, J. R., Lowe, E., and Sharples, S. (2010). The role of situation awareness for understanding signalling and control in rail operations. *Theoretical Issues in Ergonomics Science*, *11*(1–2), 84–98.

Gras, A., Moricot, C., Poirot-Delpech, S. L., and Scardigli, V. (1991). *Le pilote, le contrôleur et l'automate*. Paris: Editions de l'IRIS.

Green, D. M. and Swets, J. A. (1966). *Signal Detection Theory and Psychophysics*. London: John Wiley and Sons.

Griffin, T. G. C., Young, M. S., and Stanton, N. A. (2010). Investigating accident causation through information network modelling. *Ergonomics*, *53*(2), 198–210.

Gugerty, L. J. (1997). Situation awareness during driving: Explicit and implicit knowledge in dynamic spatial memory. *Journal of Experimental Psychology: Applied*, *3*(1), 42–66.

Gugerty, L. J. (1998). Evidence from a partial report task for forgetting in dynamic spatial memory. *Human Factors*, *40*(3), 498–508.

Guiggiani, M. (2014). *The Science of Vehicle Dynamics: Handling, Braking, and Ride of Road and Race Cars*. Berlin: Springer-Verlag.

Hall, P. M. (1981). A practical approach to road-vehicle steering installation. *Automotive Engineer* (Apr/May), 50–51.

Hamilton, I. W., Lowe, E., and Hill, C. (2007). Early route drivability assessment in support of railway investment. In R. Wilson, B. Norris, T. Clarke and A. Mills (Eds.) *People and Rail Systems: Human Factors at the Heart of the Railway*. Aldershot: Ashgate.

Hancock, P. A., Masalonis, A. J., and Parasuraman, R. (2000). On the theory of fuzzy signal detection: Theoretical and practical considerations. *Theoretical Issues in Complexity Science*, *1*(3), 207–230.

Harary, F. (1994). *Graph Theory*. Reading, MA: Addison-Wesley.

Harris, D., Chan-Pensley, J., and McGarry, S. (2005). The development of a multidimensional scale to evaluate motor vehicle dynamic qualities. *Ergonomics*, *48*(8), 964–982.

Hartman, C. H. (1978). The human factors portion of the motorcycle dynamics and handling equation. *SAE International Automotive Engineering Congress and Exposition*, Detroit, 73–78.

Hazlehurst, B., McMullen, C. K., and Gorman, P. N. (2007). Distributed cognition in the heart room: How situation awareness arises from coordinated communications during cardiac surgery. *Journal of Biomedical Informatics*, *40*(5), 539–551.

Hesse, M. and Rodrigue, J. P. (2004). The transport geography of logistics and freight distribution. *Journal of Transport Geography*, *12*(3), 171–184.

Hewett, D. G., Watson, B. M., Gallois, C., Ward, M., and Leggett, B. A. (2009). Intergroup communication between hospital doctors: Implications for quality of patient care. *Social Science and Medicine, 69*(12), 1732–1740.

Hignett, S., Carayon, P., Buckle, P., and Catchpole, K. (2013). State of science: Human factors and ergonomics in healthcare. *Ergonomics, 56*(10), 1491–1503.

Hjalmdahl, M. and Varhelyi, A. (2004). Validation of in-car observations, a method for driver assessment. *Transportation Research Part A, 38*, 127–142.

Hoffman, E. R. and Joubert, P. N. (1968). Just noticeable differences in some vehicle handling variables. *Human Factors, 10*(3), 263–272.

Hogg, D. N., Folleso, K., Strand-Volden, F., and Torralba, B. (1995). Development of a situation awareness measure to evaluate advanced alarm systems in nuclear power plant control rooms. *Ergonomics, 38*(11), 2394–2413.

Hoinville, G., Berthould, R., and Mackie, A. M. (1972). *A Study of Accident Rates amongst Motorists Who Passed or Failed an Advanced Driving Test: Report 499*. Crowthorne: Transport Road and Research Laboratory (TRRL).

Hollnagel, E. (1993). *Human Reliability Analysis–Context and Control*. London: Academic Press.

Horswill, M. S. and Coster, M. E. (2002). The effect of vehicle characteristics on drivers' risk-taking behaviour. *Ergonomics, 4*(2), 85–104.

Horswill, M. S. and McKenna, F. P. (1999). The development, validation, and application of a video-based technique for measuring an everyday risk-taking behaviour: Driver's speed choice. *Journal of Applied Psychology, 84*(6), 977–985.

Hosking, S. G., Liu, C. C., and Bayly, M. (2010). The visual search patterns and hazard responses of experienced and inexperienced motorcycle riders. *Accident Analysis and Prevention, 42*(1), 196–202.

Houghton, R. J., Baber, C., McMaster, R., Stanton, N. A., Salmon, P., Stewart, R., and Walker, G. H. (2006). Command and control in emergency services operations: A social network analysis. *Ergonomics, 49*(12–13), 1204–1225.

Howitt, D. and Cramer, D. (2016). *Research Methods in Psychology*. Cambridge, UK: Pearson.

Hurt, H. H. and DuPont, C. J. (1977). Human factors in motorcycle accidents. *SAE International Automotive Engineering Congress and Exposition*, Detroit, 54–59.

Hutchins, E. (1995a). *Cognition in the Wild*. Boston: MIT Press.

Hutchins, E. (1995b). How a cockpit remembers its speeds. *Cognitive Science, 19*(3), 265–288.

IAM. (2007). *How to Be a Better Driver: Advanced Driving - the Essential Guide*. London: Motorbooks.

IAM. (2008). 50 years of driving road safety. Available from: http://www.iam.org.uk/aboutus/history.htm [Accessed 28 May 2008].

Ingleby, J. D. (1968). *Decision-making processes in human perception and memory*. PhD thesis (unpublished), University of Cambridge.

Jacobson, M. A. I. (December 1974). Safe car handling factors. *Journal of Automotive Engineering*, 6–15.

James, N. and Patrick, J. (2004). The role of situation awareness in sport. In S. Banbury and S. Tremblay (Eds.) *A Cognitive Approach to Situation Awareness: Theory, Measures and Application* (pp. 296–316). London: Ashgate Publishers.

Jensen, R. S. (1997). The boundaries of aviation psychology, human factors, aeronautical decision making, situation awareness, and crew resource management. *International Journal of Aviation Psychology*, 7(4), 259–267.

Jones, D. G. and Endsley, M. R. (1996). Sources of situation awareness errors in aviation. *Aviation, Space, and Environmental Medicine*, 67(6), 507–512.

Joy, T. J. P. and Hartley, D. C. (1953–1954). Tyre characteristics as applicable to vehicle stability problems. *Proceedings of the Institution of Mechanical Engineers (Auto. Div.)*, 6, 113–133.

Kakimoto, T., Kamei, Y., Ohira, M., and Matsumoto, K. (2006). Social network analysis on communications for knowledge collaboration in OSS communities. *Proceedings of the 2nd International Workshop on Supporting Knowledge Collaboration in Software Development (KCSD'06)*, Tokyo, pp. 35–41.

Karl, C. A. and Bechervaise, N. E. (2003). The learning driver: Issues for provision of traveller information services. *10th World Congress and Exhibition on Intelligent Transport Systems and Services*, Madrid.

Karwowski, W. (2000). Cognitive ergonomics: Requisite compatibility, fuzziness and nonlinear dynamics. *Proceedings of the 14th Triennial Congress of International Ergonomics Association and the 35th Annual Meeting of the Human Factors Society*, San Diego, pp. 1-580–1-583.

Kattiyaportn, U. and Nel, D. (2009). The web site content of state tourism authorities analysed using Leximancer. *ANZMAC Annual Conference 2009*, Monash University, Melbourne, 27–30 November.

Kauer, M., Franz, B., Maier, A., and Bruder, R. (2015). The influence of highly automated driving on the self-perception of drivers in the context of Conduct-by-Wire. *Ergonomics*, 58(2), 321–334.

Kerr, J. S. (1991). Driving without attention mode (DWAM): A normalisation of inattentive states in driving. In A. G. Gale (Ed.) *Vision in Vehicles III*. (pp. 473–479) North Holland: Elsevier.

Klein, G. A., Calderwood, R., and MacGregor, D. (1989). Critical decision method for eliciting knowledge. *IEEE Transactions on Systems, Man, and Cybernetics*, 19(3), 462–472.

Knoll, P. M. and Kosmowski, B. B. (2002). Milestones on the way to a reconfigurable automotive instrument cluster. In J. Rutkowska, S. J. Klosowicz, and J. Zielinski (Eds.) *Proceedings of the International Society for Optical Engineering, 4759*, XIV Conference on Liquid Crystals: Chemistry, Physics, and Applications, SPIE Press, Bellingham, WA, pp. 390–394.

Lai, F., Hjalmdahl, M., Chorlton, K., and Wiklund, M. (2010). The long-term effect of intelligent speed adaptation on driver behaviour. *Applied Ergonomics*, 41(2), 179–186.

Leavitt, H. J. (1951). Some effects of certain communication patterns on group performance. *Journal of Abnormal and Social Psychology*, 46(1), 38–50.

Lechner, D. and Perrin, C. (1993). The actual use of the dynamic performances of vehicles. *Journal of Automobile Engineering*, 207(4), 249–256.

Lee, J. D. (2001). Emerging challenges in cognitive ergonomics: Managing swarms of self-organising agent-based automation. *Theoretical Issues in Ergonomics Science, 2*(3), 238–250.

Lee, J. D., Cassano-Pinché, A., and Vicente, K. J. (2005). Bibliometric analysis of human factors (1970–2000): A quantitative description of scientific impact. *Human Factors: The Journal of the Human Factors and Ergonomics Society, 47*(4), 753–766.

Lee, W., Karwowski, W., Marras, W. S., and Rodrick, D. (2003). A neuro-fuzzy model for estimating electromyographical activity of trunk muscles due to manual lifting. *Ergonomics, 46*(1), 285–309.

Lehto, M. R. and Buck, J. R. (2008). *Introduction to Human Factors and Ergonomics for Engineers.* Boca Raton: CRC Press.

Loasby, M. (1995). Is refinement and i.c.e. eroding good handling? *Automotive Engineer, 20*(1), 2–3.

Lodge, M. (1980). *Magnitude Scaling.* London: Sage.

Lowe, J. C. (1975). *The Geography of Movement.* Boston: Houghton Mifflin.

Lyons, G., Avineri, E., and Farag, S. (2008). Assessing the demand for travel information: Do we really want to know? *Proceedings of the European Transport Conference,* Noordwijkerhout, Netherlands, 6–8 October.

Ma, R. and Kaber, D. B. (2005). Situation awareness and workload in driving while using adaptive cruise control and a cell phone. *International Journal of Industrial Ergonomics, 35*(10), 939–953.

Ma, R. and Kaber, D. B. (2007). Situation awareness and driving performance in a simulated navigation task. *Ergonomics, 50*(8), 1351–1364.

MacGregor, D. G. and Slovic, P. (1989). Perception of risk in automotive systems. *Human Factors, 31*(4), 377–389.

MacMillan, N. A. and Creelman, D. C. (1991). *Detection Theory: A User's Guide.* Cambridge: Cambridge University Press.

Macquet, A. and Stanton, N. A. (2014). Do the coach and athlete have the same 'picture' of the situation? Distributed Situation Awareness in an elite sport context. *Applied Ergonomics, 45*(3), 724–733.

Magazzù, D., Comelli, M., and Marinoni, A. (2006). Are car drivers holding a motorcycle licence less responsible for motorcycle-car crash occurrence? A non-parametric approach. *Accident Analysis & Prevention, 38*(2), 365–370.

Majchrzak, A., & Borys, B. (2001). Generating testable socio-technical systems theory. *Journal of Engineering and Technology Management, 18*(3), 219–240.

Mansfield, N. J. and Griffin, M. J. (2000). Difference thresholds for automobile seat vibration. *Applied Ergonomics, 31*(3), 255–261.

Mares, P., Coyne, P., and MacDonald, B. (2013). *Roadcraft: The Police Driver's Handbook.* London: HMSO.

Marshall, S. (2005). *Streets and Patterns.* Oxon and New York: Spon Press.

Martin, N. P. D., Bishop, J. D. K., and Boies, A. M. (2017). Emissions, performance and design of UK passenger vehicles. *International Journal of Sustainable Transport, 11*(3), 230–236.

Matthews, M.D., Pleban, R.J., Endsley, M.R., and Strater, L.D. (2000). Measures of infantry situation awareness for a virtual MOUT environment. *Proceedings of the Human Performance, Situation Awareness and Automation: User Centred Design for the New Millennium Conference*, Savannah, Georgia, October 2000.

Matthews, M. L. and Cousins, L. R. (1980). The influence of vehicle type on the estimation of velocity while driving. *Ergonomics, 23*(12), 1151–1160.

May, J. L. and Gale, A. G. (1998). How did I get here? Driving without attention mode. In M. Hanson (Ed.) *Contemporary Ergonomics 1998* (pp. 456–460). London: Taylor and Francis.

McKibben, J. S. (1978). Motorcycle dynamics–fact, fiction and folklore. *SAE International Automotive Engineering Congress and Exposition*, Detroit, 63–71.

McKnight, J. A. and Adams, B. B. (1970). *Driver Education Task Analysis: Volume I—Task Descriptions*. Washington, DC: NHTSA.

McKnight, J. A. and Heywood, H. B. (1974). *Motorcycle Task Analysis*. Irvine: Motorcycle Safety Foundation.

McNicol, D. (1972). *A Primer of Signal Detection Theory*. London: George Allen & Unwin Ltd.

McRuer, D. T., Allen, R. W., Weir, D. H., and Klein, R. H. (1977). New results in driver steering control models. *Human Factors, 19*(4), 381–397.

Meriam, J. L. and Kraige, L. G. (2014). *Engineering Mechanics: Statics*, 8th edition. London: Wiley.

Metcalf, R. (1973). *Packet Communication*. MIT Project MAC Technical Report MAC TR-114.

Michon, J. A. (1985). A critical review of driver behaviour models. What do we know, what should we know? In L. Evans and R. C. Schwing (Eds.) *Human Behaviour and Traffic Safety* (pp. 485–520). New York: Plenum Press.

Michon, J. A. (1993). *Generic Intelligent Driver Support: A Comprehensive Report of GIDS*. London: Taylor and Francis.

Milliken, W. F. Jr. and Dell'Amico, F. (1968). Standards for safe handling characteristics of automobiles. In P. G. Ware (Chair) *Vehicle and Road Design for Safety, 183*, (3A), Symposium conducted by the Institution of Mechanical Engineers, London.

Monge, P. R. and Contractor, N. S. (2003). *Theories of Communication Networks*. New York: Oxford University Press.

Moray, N. (2004). Où sont les neiges d'antan? In D. A. Vincenzi, M. Mouloua, and P. A. Hancock (Eds.) *Human Performance, Situation Awareness and Automation: Current Research and Trends* (p. 4). Mahwah: LEA.

Mortimer, R. G. (1974). Foot brake pedal force capability of drivers. *Ergonomics 17*(4), 509–513.

Motor (1979). *Motor Road Test Annual 1979*. Sutton, Surrey, UK: IPC.

Mourant, R. R. and Rockwell, T. H. (1972). Strategies of visual search by novice and experienced drivers. *Human Factors, 14*(4), 325–335.

Na, X. and Cole, D.J. (2015). Game-theoretic modelling of the steering interaction between a human driver and a vehicle collision avoidance controller. *IEEE Transactions on Human-Machine Systems, 45*, 25–38.

Nakayama, T. and Suda, E. (1994). The present and future of electric power steering. *International Journal of Vehicle Design, 15*(3–5), 243–254.

Nash, C. J., Cole, D. J., and Bigler, R. S. (2016). A review of human sensory dynamics for application to models of driver steering and speed control. *Biological Cybernetics, 110*, 91–116.

Neisser, U. (1976). *Cognition and Reality: Principles and Implications of Cognitive Psychology*. San Francisco: Freeman.

Neville, T. and Salmon, P. M. (2016). Never blame the umpire – a review of situation awareness models and methods for examining the performance of officials in sport. *Ergonomics, 59*, 962–975.

Nijkamp, P., Pepping, G., and Banister, D. (1997). *Telematics and Transport Behaviour*. Berlin: Springer-Verlag.

Norman, D. A. (1981). Categorization of action slips. *Psychological Review, 88*(1), 1–15.

Norman, D. A. (1990a). *The Design of Everyday Things*, 1st edition. New York: Doubleday.

Norman, D. A. (1990b). The "problem" with automation: Inappropriate feedback and interaction, not "overautomation". *Philosophical Transactions of the Royal Society B, 327*(1241), 585–593.

Norman, D. A. (1993). *Things That Make Us Smart*. New York: Basic Books.

Norman, D. A. (2013). *The Design of Everyday Things* (revised and expanded edition). London: MIT Press (UK edition).

Nunney, M. J. (1998). *Light and Heavy Vehicle Technology*, 3rd edition. London: Newnes.

O'Hare, D., Wiggins, M., Williams, A., and Wong, W. (2000). Cognitive task analysis for decision centred design and training. In J. Annett and N. A. Stanton (Eds.) *Task Analysis* (pp. 170–190). London: Taylor and Francis.

O'Kelly, M. E. (1998). A geographer's analysis of hub-and-spoke networks. *Journal of Transport Geography, 6*(3), 171–186.

Odhams, A. M. C. and Cole, D. J. (2014). Identification of the steering control behaviour of five test subjects following a randomly curving path in a driving simulator. *International Journal of Vehicle Autonomous Systems, 12*(1), 44–64.

Ogden, G. C. (1987). Concept, knowledge and thought. *Annual Review of Psychology, 38*, 203–227.

Pacejka, H. B., and Besselink, I. J. M. (2012). *Tire and Vehicle Dynamics*. Oxford: Butterworth-Heinemann.

Patel, R. and Mohan, D. (1993). An improved motorcycle helmet design for tropical climates. *Applied Ergonomics, 24*(6), 427–431.

Patrick, J. and James, N. (2004). A task-oriented perspective of situation awareness. In S. Banbury and S. Tremblay (Eds.) *A Cognitive Approach to Situation Awareness: Theory and Application* (pp. 61–81). London: Ashgate.

Patrick, J. and Morgan, P. L. (2010). Approaches to understanding, analysing and developing situation awareness. *Theoretical Issues in Ergonomics Science, 11*(1–2), 41–57.

Pirsig, R. M. (1974). *Zen and the Art of Motorcycle Maintenance: An Inquiry into Values.* New York: William Morrow.

Pitts, S. and Wildig, A. W. (June/July 1978). Effect of steering geometry on self-centering torque and 'feel' during low-speed manoeuvres. *Automotive Engineer,* 45–48.

Plant, K. L. and Stanton, N. A. (2012). Why did the pilots shut down the wrong engine? Explaining errors in context using Schema Theory and the Perceptual Cycle Model. *Safety Science, 50*(2), 300–315.

Plant, K. L. and Stanton, N. A. (2013). The explanatory power of Schema Theory: Theoretical foundations and future applications in Ergonomics. *Ergonomics, 56*(1), 1–15.

Plant, K. L. and Stanton, N. A. (2014). All for one and one for all: Representing teams as a collection of individuals and an individual collective using a network perceptual cycle approach. *International Journal of Industrial Ergonomics, 44*(5), 777–792.

Plant, K. L. and Stanton, N. A. (2015). The process of processing: Exploring the validity of Neisser's perceptual cycle model with accounts from critical decision-making in the cockpit. *Ergonomics, 58*(6), 909–923.

Police Foundation (2007). *Roadcraft: The Essential Police Driver's Handbook.* London: Stationary Office Books.

Popp, K. and Schiehlen, W. (2010). *Ground Vehicle Dynamics.* Berlin: Springer-Verlag.

Priede, T. (1982). Road vehicle noise. In R. G. White and J. G. Walker (Eds.) *Noise and Vibration.* Chichester: Ellis Horwood.

Pugh, D. S., Hickson, D. J., Hinings, C. R., and Turner, C. (1968). Dimensions of organisation structure. *Administrative Science Quarterly, 13*(1), 65–105.

Rafferty, L. A., Stanton, N. A., and Walker, G. H. (2013). Great Expectations: A thematic analysis of situation awareness in fratricide. *Safety Science, 56,* 63–71.

Rasmussen, J., Pejtersen, A. M., and Goodstein, L. P. (1994). *Cognitive Systems Engineering.* New York: Wiley.

Ratan, P. (2013). Telematics: An emerging solution to streamline logistics. *Cargo Talk, 13*(5), 20–24.

Reason, J. (1990). *Human Error.* Cambridge: Cambridge University Press.

Reber, A. S. (1995). *The Penguin Dictionary of Psychology.* London: Penguin.

Redelmeier, D. A. and Tibshirani, R. J. (1999). Why cars in the next lane seem to go faster. *Nature, 401*(6748), 35.

Reggiani, A., Lampugnani, G., Nijkamp, P., and Pepping, G. (1995). Towards a typology of European inter-urban transport corridors for advanced transport telematics applications. *Journal of Transport Geography, 3*(1), 53–67.

Reuben, D. B., Silliman, R. A., and Traines, M. (1988). The aging driver: Medicine, policy, and ethics. *Journal of the American Geriatrics Society 36*(12), 1135–1142.

Revell, K. A. and Stanton, N. A. (2012). Models of models: Filtering and bias rings in depiction of knowledge structures and their implications for design. *Ergonomics, 55*(9), 1073–1092.

Rice, R. S. (1978). Rider skill influences on motorcycle manoeuvring. *SAE International Automotive Engineering Congress and Exposition*, Detroit, 79–90.

Ritchie, M. L., McCoy, W. K., and Welde, W. L. (1968). A study of the relation between forward velocity and lateral acceleration in curves during normal driving. *Human Factors, 10*(3), 255–258.

Robertson, S. A. and Minter, A. (1996). A study of some anthropometric characteristics of motorcycle riders. *Applied Ergonomics, 27*(4), 223–229.

Robson, G. (1997). *Cars in the UK: A Survey of All British Built and Officially Imported Cars Available in the United Kingdom since 1945: Vol 2: 1971 to 1995*. Croydon: Motor Racing Publications Ltd.

Rooney, D., Paulsen, N., Callan, V. J., Brabant, M., Gallois, C., and Jones, E. (2010). A new role for place identity in managing organizational change. *Management Communication Quarterly, 24*(1), 104–121.

Rousseau, R., Tremblay, S., and Breton, R. (2004). Defining and modelling situation awareness: A critical review. In S. Banbury and S. Tremblay (Eds.) *A Cognitive Approach to Situation Awareness: Theory and Application*. Aldershot: Ashgate.

Rumar, K. (1993). Road user needs. In A. M. Parkes and S. Franzen (Eds.), *Driving Future Vehicles* (pp. 41–48). London: Taylor and Francis.

Salas, E., Prince, C., Baker, D. P., and Shrestha, L. (1995). Situation awareness in team performance: Implications for measurement and training. *Human Factors, 37*(1), 1123–1136.

Salmon, P. M., Goode, N. A., Spiertz, A., Thomas, M., Grant, E., and Clacy, A. (2017). Is it really good to talk? Testing the impact of providing concurrent verbal protocols on driving performance. *Ergonomics, 60*(6), 770–779.

Salmon, P. M., Lenne, M. G., Walker, G. H., Stanton, N. A., and Filtness, A. (2014). Exploring schema-driven differences in situation awareness across road users: An on-road study of driver, cyclist and motorcyclist situation awareness. *Ergonomics, 57*(2), 191–209.

Salmon, P. M., Read, G. J. M., Stanton, N. A., and Lenne, M. G. 2013. The crash at Kerang: Investigating systemic and psychological factors leading to unintentional non-compliance at rail level crossings. *Accident Analysis and Prevention, 50*, 1278–1288.

Salmon, P. M. and Stanton, N. A. (2013). Situation awareness and safety: Contribution or confusion? *Safety Science, 56*, 1–5.

Salmon, P. M., Stanton, N. A., Walker, G. H., and Green D. (2006). Situation awareness measurement: A review of applicability for C4i environments. *Applied Ergonomics, 37*, 225–238.

Salmon, P. M., Stanton, N. A., Walker, G. H., and Jenkins, D. P. (2009). *Distributed Situation Awareness: Advances in Theory, Measurement and Application to Teamwork*. Aldershot: Ashgate.

Salmon, P. M., Stanton, N. A., Walker, G. H., Jenkins, D. P., Baber, C., and McMaster, R. (2008). Representing situation awareness in collaborative systems: A case study in the energy distribution domain. *Ergonomics*, *51*(3), 367–384.

Salmon, P. M., Walker, G. H., and Stanton, N. A. (2015). Broken components versus broken systems: Why it is systems not people that lose situation awareness. *Cognition, Technology and Work*, *17*(2), 179–183.

Salmon, P. M., Walker, G. H., and Stanton, N. A. (2016). Pilot error versus sociotechnical systems failure: A distributed situation awareness analysis of Air France 447. *Theoretical Issues in Ergonomics Science*, *17*(1), 64–79.

Sanders, M. S. and McCormick, E. J. (1993). *Human Factors in Engineering and Design*. London: McGraw-Hill.

Sarter, N. B. and Woods, D. D. (1991a). How in the world did I ever get into that mode: Mode error and awareness in supervisory control. *Human Factors*, *37*(1), 5–19.

Sarter, N. B. and Woods, D. D. (1991b). Situation awareness – a critical but ill-defined phenomenon. *International Journal of Aviation Psychology*, *1*(1), 45–57.

Sarter, N. B. and Woods, D. P. (1997). Team play with a powerful and independent agent: Operational experiences and automation surprises on the airbus A-320. *Human Factors*, *39*(4), 553–569.

Schiff, W. and Oldak, R. (1990). Accuracy of judging time to arrival: Effects of modality, trajectory, and gender. *Journal of Experimental Psychology: Human Perception and Performance*, *16*, 303–316.

Schulz, C. M., Endsley, M. R., Kochs, E. F., Gelb, A. W., and Wagner, A. J. (2013). Situation awareness in anaesthesia: Concept and research. *Anaesthesiology*, *118*(3), 729–742.

Schwaller, A. E. (1993). *Motor Automotive Technology*. New York: Delmar.

Sebok, A. (2000). Team performance in process control: Influence of interface design and staffing levels. *Ergonomics*, *43*(8), 1210–1236.

Seeley, T. D., Visscher, P. K., Schlegel, T., Hogan, P. M., Franks, N. R., and Marshall, J. A. (2012). Stop signals provide cross inhibition in collective decision-making by honeybee swarms. *Science*, *335*(6064), 108–111.

Segel, L. (August 1964). An investigation of automobile handling as implemented by a variable-steering automobile. *Human Factors*, *6*(4), 333–341.

Seppanen, H., Makela, J., Luokkala, P., and Virrantaus, K. (2013). Developing shared situational awareness for emergency management. *Safety Science*, *55*, 1–9.

Seppelt, B. D. and Lee, J. D. (2007). Making adaptive cruise control (ACC) limits visible. *International Journal of Human Computer Studies*, *65*, 192–205.

Setright, L. J. K. (1999). The mythology of steering feel. *Automotive Engineer*, *24*(5), 76–78.

Shinar, D., Meir, M., and Ben-Shoham, I. (1998). How automatic is manual gear shifting? *Human Factors 40*(4), 647–654.

Shu, Y. and Furuta, K. (2005). An inference method of team situation awareness based on mutual awareness. *Cognition Technology & Work*, *7*, 272–287.

Siegel, S. and Castellan, N. J. (1988). *Nonparametric Statistics for the Behavioural Sciences*, 2nd edition. London: McGraw-Hill.

Sivak, M. (1996). The information that drivers use: Is it indeed 90% visual? *Perception*, *25*(9), 1081–1089.

Smith, A. E. (May-June 2003). Automatic extraction of semantic networks from text using Leximancer. *Proceedings of HLT-NAACL*, Edmonton.

Smith, K. and Hancock, P. A. (1995). Situation awareness is adaptive, externally directed consciousness. *Human Factors*, *37*(1), 137–148.

Sneddon, A., Mearns, K., and Flin, R. (2015). Stress, fatigue, situation awareness and safety in offshore drilling crews. *Safety Science*, *56*, 80–88.

Snodgrass, J. G. and Corwin, J. (1988). Pragmatics of measuring recognition memory: Applications to dementia and amnesia. *Journal of Experimental Psychology: General*, *117*(1), 34–50.

Sontacchi, A., Frank, M., Zotter, F., Kranzler, C., and Brandl, S. (2016). Sound optimization for downsized engines. In: A. Fuchs, E. Nijman, and H. H. Priebsch (Eds.) *Automotive NVH Technology* (pp. 13–28). SpringerBriefs in Applied Sciences and Technology. Cham, Switzerland: Springer.

Sorensen, L. J. and Stanton, N. A. (2013). Y is best: How distributed situational awareness is mediated by organisational structure and correlated with task success. *Safety Science*, *56*, 72–79.

Sorensen, L. J. and Stanton, N. A. (2015). Exploring compatible and incompatible transactions in teams. *Cognition, Technology and Work*, *17*(3), 367–380.

Sorensen, L. J. and Stanton, N. A. (2016). Inter-rater reliability and content validity of network analysis as a method for measuring distributed situation awareness. *Theoretical Issues in Ergonomics Science*, *17*(1), 42–63.

Sorensen, L. J., Stanton, N. A., and Banks, A. P. (2011). Back to SA school: Contrasting three approaches to situation awareness in the cockpit. *Theoretical Issues in Ergonomics Science*, *12*(6), 451–471.

Southall, D. (1985). The discrimination of clutch-pedal resistances. *Ergonomics*, *28*(9), 1311–1317.

Spath, D., Braun, M., and Hagenmeyer, L. (2006). Human factors and ergonomics in manufacturing and process control. In G. Salvendy (Ed.) *Handbook of Human Factors and Ergonomics*, (pp. 1597–1626). Hoboken: John Wiley & Sons.

Stammers, R. B. and Astley, J. A. (1987). Hierarchical task analysis: Twenty years on. In E. D. Megaw (Ed.), *Contemporary Ergonomics 1987* (pp. 135–139). Proceedings of the Ergonomics Society's 1987 Annual Conference, April 6-10, 1987, Swansea, Wales. London: Taylor and Francis.

Stanton, N. A. (2006). Hierarchical task analysis: Developments, applications and extensions. *Applied Ergonomics*, *37*(5), 55–79. Invited paper to special issue on 'Fundamental Reviews of Ergonomics'.

Stanton, N. A. (2010). Situation awareness: Where have we been, where are we now and where are we going? *Theoretical Issues in Ergonomics Science*, *11*(1), 1–6.

Stanton, N. A. (2014). Representing distributed cognition in complex systems: How a submarine returns to periscope depth. *Ergonomics*, *57*(3), 403–418.

Stanton, N. A. (2016). Distributed situation awareness. *Theoretical Issues in Ergonomics Science*, *17*(1), 1–7.

Stanton, N. A., Baber, C., Walker, G. H., Salmon, P., and Green, D. (2004). Toward a theory of agent-based systemic situational awareness. In D. A. Vincenzi, M. Mouloua, and P. A. Hancock (Eds.) *Proceedings of the Second Human Performance, Situation Awareness and Automation Conference (HPSAAII)*, Daytona Beach, FL, 22–25 March.

Stanton, N. A., Chambers, P. R. G., and Piggott, J. (2001). Situational awareness and safety. *Safety Science*, *39*(3), 189–204.

Stanton, N. A. and Marsden, P. (1996). From fly-by-wire to drive-by-wire: Safety implications of automation in vehicles. *Safety Science*, *24*(1), 35–49.

Stanton, N. A. and Pinto, M. (2000). Behavioural compensation by drivers of a simulator when using a vision enhancement system. *Ergonomics*, *43*(9), 1359–1370.

Stanton, N. A., Salmon, P. M., Rafferty, L., Walker, G. H., Baber, C., and Jenkins, D. P. (2013). *Human Factors Methods: A Practical Guide for Engineering and Design*, 2nd edition. Farnham, UK: Ashgate.

Stanton, N. A., Salmon, P. M., and Walker, G. H. (2015). Let the reader decide: A paradigm shift for situation awareness in sociotechnical systems. *Journal of Cognitive Engineering and Decision Making*, *9*(1), 44–50.

Stanton, N. A., Salmon, P. M., Walker, G. H., and Jenkins, D. P. (2009a). Genotype and phenotype schemata as models of situation awareness in dynamic command and control teams. *International Journal of Industrial Ergonomics*, *39*(3), 480–489.

Stanton, N. A., Salmon, P. M., Walker, G. H., and Jenkins, D. P. (2009b). Genotype and phenotype schemata and their role in distributed situation awareness in collaborative systems. *Theoretical Issues in Ergonomics Science*, *10*(1), 43–68.

Stanton, N. A., Salmon, P. M., Walker, G. H., Baber, C., and Jenkins, D. P. (2005). *Human Factors Methods: A Practical Guide for Engineering and Design*. Aldershot: Ashgate.

Stanton, N. A., Salmon, P. M., Walker, G. H., Hancock, P. A., and Salas, E. (2017). State-of-science: Situation awareness in individuals, teams and systems. *Ergonomics*, *60*(4), 449–466.

Stanton, N. A., Salmon, P. M., Walker, G. H., and Jenkins, D. P. (2010). Is situation awareness all in the mind? *Theoretical Issues in Ergonomics Science*, *11*(1), 29–40.

Stanton, N. A., Stewart, R., Harris, D., Houghton, R. J., Baber, C., McMaster, R., Salmon, P., Hoyle, G., Walker, G., Young, M. S., Linsell, M., Dymott, R., and Green, D. (2006). Distributed situation awareness in dynamic systems: Theoretical development and application of an ergonomics methodology. *Ergonomics*, *49*(12–13), 1288–1311.

Stanton, N. A. and Walker, G. H. (2011). Exploring the psychological factors involved in the Ladbroke Grove rail accident. *Accident Analysis and Prevention*, *43*(3), 1117–1127.

Stanton, N. A., Walker, G. H., Young, M. S., Kazi, T. A., and Salmon, P. (2007). Changing drivers' minds: The evaluation of an advanced driver coaching system. *Ergonomics. Special Issue on Driver Safety*, *50*(8), 1209–1234.

Stanton, N. A. and Young, M. S. (1998). Vehicle automation and driving performance. *Ergonomics, 41*(7), 1014–1028.

Stanton, N. A. and Young, M. S. (1999). *A Guide to Methodology in Ergonomics.* London: Taylor and Francis.

Stanton, N. A. and Young, M. S. (2000). A proposed psychological model of driving automation. *Theoretical Issues in Ergonomics Science, 1,* 315–331.

Stanton, N. A. and Young, M. S. (2005). Driver behaviour with adaptive cruise control. *Ergonomics, 48*(10), 1294–1313.

Stanton, N. A., Young, M. S., and McCaulder, B. (1997). Drive-by-wire: The case of mental workload and the ability of the driver to reclaim control. *Safety Science, 27*(2–3), 149–159.

Stanton, N. A., Young, M. S., and Walker, G. H. (2007). The psychology of driving automation: A discussion with Professor Don Norman. *International Journal of Vehicle Design, 45*(3), 289–306.

Stanton, N. A., Young, M. S., Walker, G. H., Turner, H., and Randle, S. (2001). Automating the driver's control tasks. *International Journal of Cognitive Ergonomics, 5*(3), 221–236.

Stevens, N., Salmon, P. M., Walker, G. H., and Stanton, N. A. (2016). Off the beaten track: Situation awareness in experienced and novice off-road drivers. *ARSC2016 Australasian Road Safety Conference.*

Stewart, R., Stanton, N. A., Harris, D., Baber, C., Salmon, P. M., Mock, M., Tatlock, K., Wells, L., and Kay, A. (2008). Distributed situation awareness in an airborne warning and control system: Application of novel ergonomics methodology. *Cognition Technology and Work, 10*(3), 221–229.

Sukthankar, R. (1997). *Situational awareness for tactical driving* Doctoral dissertation (unpublished), Carnegie Mellon University.

Summala, H. (1996). Accident risk and driver behaviour. *Safety Science, 22*(1–3), 103–117.

Summala, H. (2000). Automatization, automation, and modelling of driver's behaviour. *Recherche, Transports, Securité (RTS), 66*(1), 34–45.

Taylor, R. M. (1990). Situational awareness rating technique (SART): The development of a tool for aircrew systems design. *Situational Awareness in Aerospace Operations (AGARD-CP-478)*, 3/1–3/17, Neuilly-sur-Seine: NATO-AGARD.

Taylor, R. M. and Selcon, S. J. (1994). Subjective measurement of situation awareness. In Y. Queinnec and F. Danniellou (Eds.) *Designing for Everyone: Proceedings of the 11th Congress of the International Ergonomics Association* (pp. 789–791). London: Taylor and Francis.

Thomson, J. M. (1977). *Great Cities and Their Traffic.* London: Victor Gollancz.

Thurstone, L. L. (1928). Attitudes can be measured. *American Journal of Sociology, 33*(4), 529–544.

Tijerina, L., Gleckler, M., Stoltzfus, D., Johnstone, S., Goodman, M. J., and Wierwille, W. W. (1998). *A Preliminary Assessment of Algorithms for Drowsy and Inattentive Driver Detection on the Road.* Washington, DC: NHTSA.

Underwood, G., Chapman, P., Brocklehurst, N., Underwood, J., and Crundall, D. (2003). Visual attention while driving: Sequences of eye fixations made by experienced and novice drivers. *Ergonomics, 46*(6), 629–646.

Van Winsen, R. and Dekker, S. W. A. (2015). SA Anno 1995: A commitment to the 17th century. *Journal of Cognitive Engineering and Decision Making, 9*(1), 51–54.

Vicente, K. J. (1999). *Cognitive Work Analysis: Toward Safe, Productive, and Healthy Computer—Based Work.* Mahwah: Lawrence Erlbaum Associates.

Walker, G. H. (2004). Verbal protocol analysis. In N. A. Stanton (Eds.) *Handbook of Human Factors and Ergonomics Methods* (pp. 30-1–30-7). Boca Raton: CRC Press.

Walker, G. H. (February 2008). The disbenefits of technology advances. Raising awareness: How modern car design affects drivers. *Traffic Engineering and Control.*

Walker, G. H. (2016). Fortune favours the bold. *Theoretical Issues in Ergonomics Science, 17*(4), 452–458.

Walker, G. H., Salmon, P. M., Bedinger, M., and Stanton, N. A. (2017). Quantum ergonomics: Shifting the paradigm of the systems agenda. *Ergonomics: Special Issue on New Paradigms, 60*(2), 157–166.

Walker, G. H., Stanton, N. A., and Chowdhury, I. (2013). Situational awareness and self explaining roads. *Safety Science, 56*, 18–28.

Walker, G. H., Stanton, N. A., Kazi, T. A., Salmon, P. M., and Jenkins, D. P. (2009). Does advanced driver training improve situation awareness? *Applied Ergonomics, 40*(4), 678–687.

Walker, G. H., Stanton, N. A., and Salmon, P. M. (2015). *Human Factors in Automotive Engineering and Technology.* Boca Raton: CRC Press.

Walker, G. H., Stanton, N. A., Salmon, P. M., Jenkins, D. P., and Rafferty, L. (2010). Translating concepts of complexity to the field of ergonomics. *Ergonomics, 53*(10), 1175–1186.

Walker, G. H., Stanton, N. A., and Young, M. S. (2001a). An on-road investigation of vehicle feedback and its role in driver cognition: Implications for cognitive ergonomics. *International Journal of Cognitive Ergonomics, 5*(4), 421–444.

Walker, G. H., Stanton, N. A., and Young, M. S. (2001b). An on-road investigation of feedback and cognitive processing in naturalistic driving. In J. Smith, G. Salvendy, and Kasdorf, M. R. (Eds.) *HCI International 2001; 9th International Conference on Human-Computer Interaction,* LEA, New Jersey.

Walker, G. H., Stanton, N. A., and Young, M. S. (2001c). Hierarchical task analysis of driving: A new research tool. In M. A. Hanson (Ed.) *Contemporary Ergonomics 2001* (pp. 435–440). London: Taylor and Francis.

Walker, G. H., Stanton, N. A., and Young, M. S. (2001d). Where is computing driving cars? A technology trajectory of vehicle design. *International Journal of Human Computer Interaction, 13*(2), 203–229.

Walker, G. H., Stanton, N. A., and Young, M. S. (2006). The ironies of vehicle feedback in car design. *Ergonomics, 49*(2), 161–179.

Walker, G. H., Stanton, N. A., and Young, M. S. (2007). What's happened to car design? An exploratory study into the effect of 15 years of progress on driver situation awareness. *International Journal of Vehicle Design, 45*(1/2), 266–282.

Walker, G. H., Stanton, N. A., and Young, M. S. (2008). Feedback and driver situation awareness (SA): A comparison of SA measures and contexts. *Transportation Research Part F: Traffic Psychology and Behaviour, 11*(4), 282–299.

Walton, D. and Bathurst, J. (1998). An exploration of the perception of the average driver's speed compared to perceived driver safety and driving skill. *Accident Analysis and Prevention, 30*(6), 821–830.

Warning over design of 'silent killer cars'. *Manchester Evening News*, 15 October, 2007.

Watts, D. J. and Strogatz, S. H. (1998). Collective dynamics of 'small-world' networks. *Nature, 393*(4), 440–442.

Weber, R. P. (1990). *Basic Content Analysis*. London: Sage Publications.

Weisstein, E. W. (2008). Graph diameter. From MathWorld—A Wolfram Web Resource. Available from: http://mathworld.wolfram.com/GraphDiameter.html [Accessed 17 September 2008]

Welford, A. T. (1968). *Fundamentals of Skill*. London: Methuen.

Wellens, A. R. (1993). Group situation awareness and distributed decision making: From military to civilian applications. In N. J. Castellan (Ed.) *Individual and Group Decision Making: Current Issues* (pp. 267–287). Mahwah: Erlbaum Associates.

West, R., Elander, J., and French, D. (1992). *Decision Making, Personality and Driving Style as Correlates of Individual Accident Risk: Contractor Report 309*. Crowthorne: Transport Research Laboratory.

Wickens, C. D. (1992). *Engineering Psychology and Human Performance*. New York: Harper Collins.

Wickens, C. D. (2008). Situation awareness: Review of Mica Endsley's 1995 articles on situation awareness theory and measurement. *Human Factors, 50*(3), 397–403.

Wilson, J. R. (2012). Fundamentals of systems ergonomics. *Work, 41*, 3861–3868.

Wilson, J. R. and Rajan, J. A. (1995). Human-machine interfaces for systems control. In J. R. Wilson and E. N. Corlett (Eds.) *Evaluation of Human Work: A Practical Ergonomics Methodology* (pp. 357–405). London: Taylor and Francis.

Winsum, W. M. and Godthelp, H. (1996). Speed choice and steering behavior in curve driving. *Human Factors, 38*(3), 434–441.

Woods, D. D. (1988). Coping with complexity: The psychology of human behaviour in complex systems. In L. P. Goodstein, H. B. Andersen, and S. E. Olsen (Eds.) *Tasks, Errors and Mental Models*. London: Taylor and Francis.

Woods, D. D. and Dekker, S. (2000). Anticipating the effects of technological change: A new era of dynamics for human factors. *Theoretical Issues in Ergonomics Science, 1*(3), 272–282.

Woods, D. D., Johannesen, L. J., Cook, R. I., and Sarter, N. B. (1994). *Behind Human Error: Cognitive Systems, Computers and Hindsight*. Columbus, OH: CSERIAC.

Wu, J. D., Lee, T. H., and Bai, M. R. (2003). Background noise cancellation for hands-free communication system of car cabin using adaptive feedforward algorithms. *International Journal of Vehicle Design, 31*(4), 440–451.

Young, M. S. and Stanton, N. A. (2001). Mental workload: Theory, measurement and application. In W. Karwowski (Ed.) *International Encyclopedia of Ergonomics and Human Factors* (2nd ed., *Vol. 1*, pp. 507–509). London: Taylor and Francis.

Young, M. S. and Stanton, N. A. (2002a). Malleable attentional resources theory: A new explanation for the effects of mental underload on performance. *Human Factors, 44*(3), 365–375.

Young, M. S. and Stanton, N. A. (2002b). Attention and automation: New perspectives on mental underload and performance. *Theoretical Issues in Ergonomics Science, 3*(2), 178–194.

Young, M. S. and Stanton, N. A. (2004). Taking the load off: Investigations of how adaptive cruise control affects mental workload. *Ergonomics, 47*(9), 1014–1035.

Young, M. S. and Stanton, N. A. (2006). Mental workload: Theory, measurement and application. In W. Karwowski (Ed.) *International Encyclopedia of Ergonomics and Human Factors* (2nd ed., *Vol. 1*, pp. 818–821). London: Taylor and Francis.

Young, M. S. and Stanton, N. A. (2007). What's skill got to do with it? Vehicle automation and driver mental workload. *Ergonomics, 50*(8), 1324–1339.

Zuboff, S. (1988). *In the Age of the Smart Machine: The Future of Work and Power.* London: Heinemann.

Index

D

CPSIA information can be obtained
at www.ICGtesting.com
Printed in the USA
LVHW06*2255181018
594104LV00008B/97/P